Polarised Light in Science and Nature

Professor David Pye, born in 1932, was educated at Queen Elizabeth's Grammar School, Mansfield, University College of Wales, Aberystwyth and Bedford College for Women, London. He was lecturer and then reader at King's College and has been Professor of Zoology at Queen Mary, University of London since 1973. He developed an early fascination for bat 'radar' and the electronic instrumentation necessary for the study of animal ultrasound. He was a Founder Director in 1976 of QMC Instruments Ltd, which produced large numbers of commercial ultrasound detectors, mainly for biological studies. He has travelled widely in order to study tropical bats and latterly has developed an interest in ultraviolet light and polarisation in the visual world of animals. A strong supporter of demonstration lectures, he gave the Royal Institution Christmas Lectures in 1985, and shares the Dodo's opinion that 'the best way to explain it is to do it'. This book arose from a demonstration lecture which he calls 'Polar Explorations—in Light'.

Polarised Light in Science and Nature

David Pye

Emeritus Professor
Queen Mary, University of London

IoP

Institute of Physics Publishing
Bristol and Philadelphia

British Library Cataloguing-in-Publication Data

A catalogue record for this book is available from the British Library.

ISBN 0 7503 0673 4

Library of Congress Cataloging-in-Publication Data are available

Commissioning Editor: John Navas
Production Editor: Simon Laurenson
Production Control: Sarah Plenty
Cover Design: Victoria Le Billon
Marketing Executive: Colin Fenton

Published by Institute of Physics Publishing, wholly owned by The Institute of Physics, London

Institute of Physics Publishing, Dirac House, Temple Back, Bristol BS1 6BE, UK

US Office: Institute of Physics Publishing, The Public Ledger Building, Suite 1035, 150 South Independence Mall West, Philadelphia, PA 19106, USA

Typeset in TEX using the IOP Bookmaker Macros
Printed in the UK by Hobbs the Printers, Totton, Hampshire

Contents

Preface

We humans cannot see when light is polarised and this leads us to unfortunate misapprehensions about it. Even scientists who should know better, often assume that polarised light is an obscure topic of specialised interest in only a few rather isolated areas; in fact it is a universal feature of our world and most of the natural light that we see is at least partially polarised. In the Animal Kingdom, insects and many other animals exploit such natural polarisation in some fascinating ways since they do not share this human limitation and can both detect and analyse polarisation. It may be our unfamiliarity with this aspect of light that also makes many people think it is a 'difficult' subject, yet the basis is extremely simple. When such misconceptions are overcome, the phenomena associated with polarisation are found to be important throughout science and technology—in natural history, and biology, geology and mineralogy, chemistry, biochemistry and pharmacology, physics and astronomy and several branches of engineering, including structural design, communications, high speed photography and sugar refining, as well as crafts such as glassblowing and jewellery. They also involve some very beautiful effects, most of which are easy to demonstrate and manipulate.

Our general unawareness of what we are missing is indeed a great pity. This book hopes to put all this right and enrich its readers' perception of the world. A small degree of repetition and overlap has seemed necessary in order to make each topic complete; I hope it does not become trying. The text deliberately uses no maths and only the minimum of technical terms—it is hoped that rejecting jargon, however precise and convenient it may be to the specialist, will make the stories more accessible to the newcomer. In any case, the book covers such a wide range of science that each chapter would need a separate vocabulary to be introduced and defined, which would become tedious and might well deter many readers. Descriptive terms or even circumlocutions are sometimes quicker in the end. In any case this is not a textbook; it does not aim to help directly with any particular course of study but is essentially interdisciplinary, hoping to interest any enquiring mind: a reader taking any course or none at all. Such cross-cultural influences appear to be deplorably unfashionable at present and this volume hopes to defend them by dealing with some simple unifying principles.

The book grew from a demonstration lecture, called 'Polar Explorations in Light' that I first developed for young audiences, initially at the Royal Institution of Great Britain. The 1874 classic book on polarised light by William Spottiswood also developed from a series of public lectures and I only hope that following such illustrious footsteps will achieve similar success. My own lecture has expanded to become a show that can now be adapted to almost any kind of audience. I was greatly drawn to the subject precisely because it brings in such a wide variety of phenomena across science, and because it allows one to perform some extremely beautiful demonstrations that never fail to elicit satisfying reactions from audiences of any age. It was gratifying, therefore, when the publishers suggested the possibility of a derivative book. I have tried to retain an element of the demonstration approach and, although no actual do-it-yourself-at-home recipes are given, I hope the descriptions are sufficiently helpful (and stimulating) to enable any resourceful reader to try things out. It is very rewarding to do and often quite easy, while many of the effects are much more beautiful than can be shown in photographs. Polaroid, as described in chapter 1, is widely available but if the larger sizes of sheet seem a little expensive, then the reflecting polarisers described in chapter 7 allow much to be done with the expenditure of nothing but a little ingenuity.

A reading list has been included in the hope that readers will want to find out more about some of the fields introduced here. This book does not attempt to be comprehensive in its treatment, simply to attract and intrigue. As always there is much to learn about a topic once you begin to get into it.

Acknowledgments

Several colleagues from Queen Mary, University of London have helped me to develop some of the demonstrations used in the lectures. Ray Crundwell (Media Services) was solely responsible for processing the photographs presented here and gave much invaluable advice. Others who have been especially helpful and have contributed in many different ways to the emergence of this book include Isaac Abrahams, Gerry Moss and Stuart Adams (Chemistry), Bill French and Kevin Schrapel (Earth Sciences), Edward Oliver (Geography), John Cowley (Glass Workshop), David Bacon (Media Services) and Linda Humphreys and Lorna Mitchell (Library). Much encouragement and/or material help have been generously provided by Sir Michael Berry, Ken Edwards, Ilya Eigenbrot, Cyril Isenberg, Mick Flinn, Ken Sharples (Sharples Stress Engineering Ltd), Frank James and Bipin Parma (the Royal Institution of Great Britain), Dick Vane-Wright and Malcolm Kerley (Entomology Department, Natural History Museum), Chirotech Technology Ltd, Abercrombie and Kent Travel, Ernst Schudel (Photo-Suisse, Grindelwald, Switzerland), Murray Cockman (Atomic Weapons Research Establishment), Michael Downs (National Physical Laboratory), Jørgen Jensen (Skodsborg, Denmark), Søren Thirslund (Helsingor, Denmark), Hillar Aben (Estonian Academy of Science, Tallinn) and Brian Griffin (Optical Filters Ltd). The British Library, the Linnean Society Library, the Royal Society Library and Marie Odile Josephson of the Cultural Service at the French Embassy in London have all been enormously helpful, especially in tracing historical details.

Chapter 1

Aligning the waves

Polarised light is quite simply light in which the waves are all vibrating in one fixed direction. Most waves (sound waves are an exception) involve a vibration at right angles to their path. Waves on water go only up and down but the waves on a wiggled rope can be made to go up and down or from side to side or in any other direction around their line of travel. In just the same way, light waves can vibrate in any direction across their path. Now in 'ordinary' unpolarised light the direction of vibration is fluctuating rapidly, on a time scale of about 10^{-8} s (a hundredth of a millionth of a second), and randomly through all possible directions around the path of the ray. Polarisation simply consists of forcing the waves to vibrate in a single, constant direction. A number of simple methods for showing that light is polarised and determining the direction of vibration will be described in this book, especially in chapters 2, 3 and 7.

An analogy with polarised light can be made by a wiggled rope that is passed through a narrow slit such as a vertical gap between fence posts or railings (figure 1.1). Vertical wiggles will pass unhindered through the slit but horizontal waves will be reduced or completely suppressed. If the rope is wiggled in all directions randomly, only the vertical components will pass through the slit. The equivalent effect with electromagnetic waves can be demonstrated with a low power microwave generator and detector (figure 1.2). Such waves, at a wavelength of 3 cm, are similar to those used in a microwave oven but in this case at less than a hundred-thousandth of the power of an oven. Because of the way it works, the generator produces waves that vibrate in one direction only—polarised waves—and the detector is only sensitive to waves polarised in one

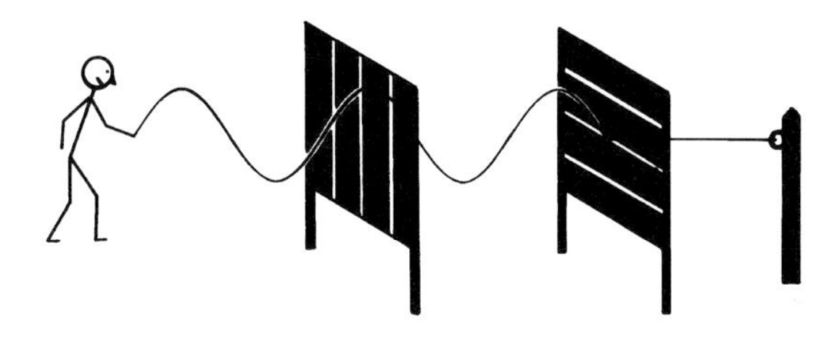

Figure 1.1. Waves pass along a wiggled rope. Where the rope passes through a slit in a fence, the waves continue if they are aligned with the slit but are stopped if they are transverse to the slit.

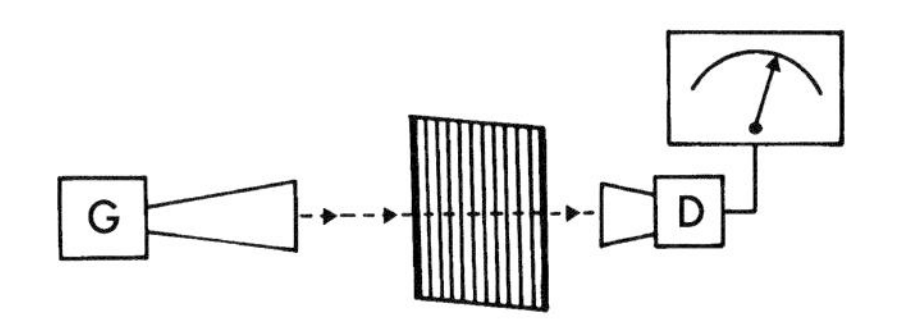

Figure 1.2. Apparatus to demonstrate polarisation with microwaves. A generator produces electromagnetic microwaves (3 cm wavelength radio waves) that vibrate in one direction only. A tuned receiver detects these waves only if they vibrate in one direction as shown by the deflection on a meter dial. With the two devices aligned, the meter is deflected, but detection ceases when either of them is twisted by 90° around their common axis. A grid with spacings less than the wavelength allows the waves to pass in one orientation but blocks them when it is turned onto its side around the axis of the beam.

direction. When the two are aligned, facing one another, the waves are detected as shown by the needle of a meter attached to the receiver, but if either unit is rotated onto its side, then reception ceases and the meter returns to zero although the waves are still being propagated.

With the generator and detector realigned and a signal being received, a wire grid with a spacing of about 7–8 mm (roughly one-quarter of a wavelength) can be held across the beam. When the wires are in line with the direction of vibration, the beam is completely blocked, but rotating the grid by 90° restores full reception and the grid becomes completely 'transparent'. (A grid aligned with the direction of vibration reflects the waves away, so blocking their path although one might expect this to be the orientation that allows them through.) It is easy to imagine that if the direction of vibration of the waves fluctuated randomly, then the grid would block all the components with one direction and pass the rest, all vibrating in the other direction at right angles to the grid wires.

To be strict, these waves are known to consist of a vibration of the electrical field at right angles to an associated vibration of the magnetic field, hence the name electromagnetic waves. So there are actually two directions of vibration in any given wave. Most scientists and engineers assume 'the' vibration to be the electrical one and simply remember that the magnetic effect is there too, at right angles. Traditionally physicists did it the other way round, with 'the' vibration being the magnetic one, but nowadays this seems to be changing. Nevertheless one needs to check what convention any particular author is using. In common with *almost* universal current practice, this book refers to 'the' direction of vibration as that of the electric component. (Earlier texts referred to the 'plane' of polarisation and to 'plane polarised' light; for several good reasons these terms are now better replaced, as in this book, by the 'direction' of polarisation and 'linearly polarised' light respectively.)

Light waves are also electromagnetic waves, with exactly the same nature as microwaves except that the wavelength is about fifty thousand times smaller. The equivalent of wire grid polarisers can be made by embedding very fine arrays of parallel metallic whiskers in a thin transparent film; these are used at the rather longer infrared wavelengths and have also been made to work for light. But in general the short wavelengths of light require one to look for structures on the scale of atoms and molecules. Early studies of polarisation used crystals whose regular lattice of atoms can interact with light waves in some interesting ways, as described in chapter 3. Such devices were tricky to make and therefore expensive. They were also quite long and narrow, with a small area, or working aperture, or else they were of poor optical quality, which limited their use in optical instruments. In 1852 William Bird Herapath described a way of making thin crystals with strong polarising properties from a solution of iodine and quinine sulphate. Unfortunately these crystals, which came to be called herapathite, were so extremely delicate that their application was seldom practical, although Sir David Brewster did try some in his kaleidoscopes (see chapter 2).

Then, around 1930, Edwin Land developed ways of aligning microscopic crystals of herapathite while fixing them as a layer on a plastic sheet to make a thin, rugged polarising film that was soon called J-type polaroid. A series of developments followed rapidly and soon superseded the original material. H-type polaroid was made by absorbing iodine on a stretched sheet of polyvinyl alcohol. K-type

Figure 1.3. One polariser, a sheet of polaroid film, only allows half of the random, unpolarised light to pass but this is then all vibrating in one direction. Such polarised light passes easily through a second polaroid that is aligned with the first (left) but is completely blocked when the two polaroids are crossed (right). This simple but striking demonstration can easily be demonstrated to an audience on an overhead projector.

polaroid was made without iodine by stretching polyvinyl alcohol films and then dehydrating them. In a sense these materials resemble the wire grid used with microwaves, since the long polymer molecules are aligned by the stretching process. Both H-type and K-type are still much used, sometimes combined as HR-type polaroid which is effective for infrared waves. By adding dyes to the material, L-type polarisers were created that only polarised a part of the spectrum while freely transmitting the rest or, conversely, that transmitted only one colour. These materials soon found a very wide range of applications from components in scientific instruments to domestic sunglasses. Land always hoped that polaroid filters, with the direction of vibration set at 45°, would become standard on car headlamps. Crossed polaroids in the windscreen or on glasses worn by the driver would then block the glare of oncoming traffic while being aligned with the car's own lamps, so making only a small reduction in their effectiveness and even allowing them to remain undipped. Clearly this would only be helpful if every vehicle were so equipped and it has not come about.

The availability of polaroid has made observations of polarised light enormously more accessible as well as greatly increasing the applications of polarised light. For the highest optical quality, professionals still sometimes need to use expensive and inconvenient

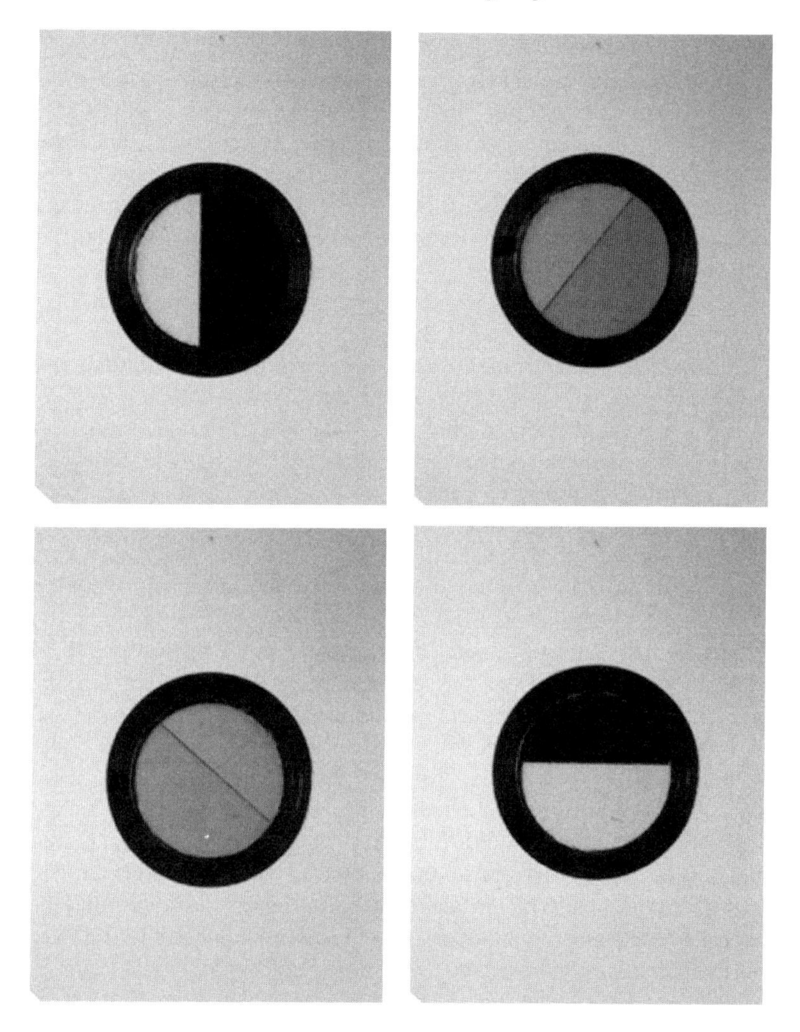

Figure 1.4. A simple polariscope to detect polarisation can be made by two pieces of polaroid film with their polarisation directions at right angles to each other. When it is rotated against a background of polarised light, each half turns dark in turn but at the precise intermediate positions they are equally 'grey'. Except in this latter state, the contrast between the two pieces is a more sensitive indicator than can be achieved by rotating a single piece to see if darkening occurs. An alternative arrangement with the polaroids at right angles is shown in colour plate 7.

Nicol prisms (see chapter 3) but polaroid is generally cheap, robust, thin and can be easily cut to any desired shape. It can be incorporated into cameras, microscopes and other instruments without any radical redesign or machining and it allows any amateur tremendous scope for exploiting the many properties of polarised light, which would have been inconceivable even to the specialist before 1930. The main disadvantage with polaroid is that, because it absorbs half the energy of the light, it can easily get very hot, especially if infrared 'heat-rays' are involved as with powerful filament lamps. It may be necessary to use a heat filter and/or a cooling fan in some cases.

A simple demonstration of the polarising action of polaroid, and also a test that it is polaroid rather then a simple tinted filter, is to overlap two pieces and rotate one (figure 1.3). When the polaroids are aligned, the light that passes through the first is also passed by the other. But when they are crossed, almost no light passes through both—they look black where they overlap. With tinted filters, of course, two always look darker than one and rotation makes no difference. The direction of polarisation for any given specimen of polaroid can easily be determined by looking through it at light reflected from a horizontal shiny surface such as gloss paint, varnish, water or glass. Such light is horizontally polarised, as described in detail in chapter 7. So when the polariser is turned to the vertical, the reflection appears to be dimmed or completely suppressed. A small mark can then be made in one corner of the polariser for future reference.

An instrument used to detect the presence of polarisation is called a polariscope. In its simplest form it is just a piece of polaroid or any other polariser that is rotated as a source of light is viewed through it. If the brightness of the source appears to vary with the rotation, then the light must be polarised. But this is often tedious and a slow fluctuation in brightness is not always easy to judge. It is much better to have two pieces of polaroid orientated at right angles and placed next to each other. A contrast in brightness can then be seen quickly and much more sensitively. If the direction of polarisation happens to be at exactly 45° to the two polariser directions then they will appear equally bright (figure 1.4) but this is unlikely to occur often and is easily eliminated by rocking the instrument slightly around its axis. Even better polariscopes will be described in the next chapter.

Chapter 2

Changing direction

Interesting things start to happen when polarised light passes through cellophane. A simple jam-pot cover, obtainable in packets of 20 from a newsagent, can rotate the direction of polarisation by 90°. One such film, placed between crossed polarisers, can twist the direction of vibration of light from the first polariser so that it then passes freely through the second. It thus appears as a clear, circular 'window' through the darkened background—the effect is especially striking when done on an overhead projector (figure 2.1). But this only happens with certain orientations of the disc, for turning it makes the 'window' darken and brighten four times during each rotation. The explanation for this depends on a property of the film called birefringence.

Cellophane is a polymer formed by the joining together of glucose molecules in long chains, and to make a thin film the material is extruded under pressure through a narrow slot so that the polymer chains become aligned. Now light vibrating in a direction parallel with the polymer chains propagates through the film at a different speed from light vibrating at right angles, across the polymer chains. The speed of light in any material is responsible for the refraction or bending of the rays when entering or leaving, and is indicated by its refractive index, or its 'refringence'. So a material with two speeds of light, depending on the direction of polarisation, must have two refractive indices and is said to be birefringent. In a thin film of cellophane, the two different angles of refraction are not noticeable but the two speeds can have a profound influence on polarisation.

[Some readers may prefer to skip the next two paragraphs although the argument is well worth following as it may dispel

Figure 2.1. A jam-pot cover placed between crossed polaroids may form a bright, clear 'window' by twisting the polarisation direction through 90°. Rotating the cover by 45° in either direction 'closes the window'.

much of the 'mystery' often associated with polarised light; but the aesthetic effects that follow can be enjoyed without necessarily tackling the theory of their origins.]

The direction of polarisation of a wave can be simply represented as an arrow, like a hand on the face of a clock, that shows the direction of vibration as seen when the light is approaching (figure 2.2). If the length of the arrow is then made proportional to the amplitude of the wave, it is called a vector. Now any vector can be considered to be equivalent to two other vectors with any two directions and lengths; if the pair of arrows are used to make two sides of a parallelogram, then the diagonal between them is the equivalent single vector or resultant (figure 2.2). They are just like a parallelogram of forces and indeed they do actually represent the forces of the electrical field associated with the light wave.

When the direction of polarisation is either parallel with the polymer chains or at right angles to them, then the light is unaffected, apart from a slight delay in traversing the film. But when the direction of polarisation is at 45° to the polymer chains, it is divided into two components that traverse the material with slightly different delays, so that on emergence one component is retarded with respect to the other. Now a jam-pot cover gives a relative retardation of just half a wavelength, so that it acts as what is called a half-wave plate. In the retarded wave, the peaks come where the troughs would have been (and vice versa) so the vector arrow is effectively

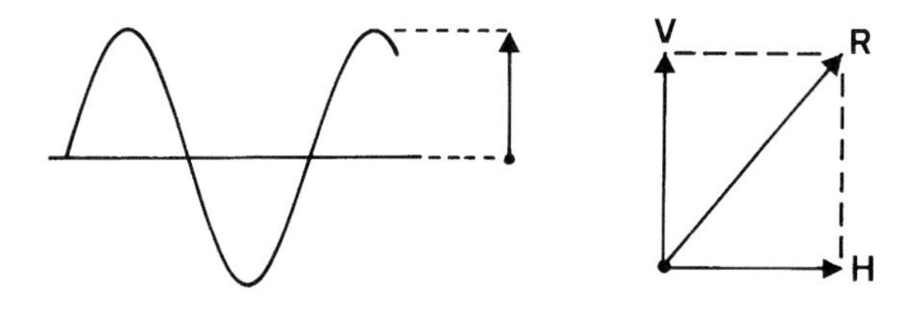

Figure 2.2. Left: a simple wave can be represented by an arrow called a vector whose direction indicates the direction of the vibration and whose length indicates its strength or amplitude. It is as if the wave is viewed as it approaches, and only the height and direction of the peaks are shown. Right: a vertically polarised wave (V) and a slightly weaker horizontally polarised wave (H) are together equivalent to the resultant vector (R), as shown by completing the parallelogram and its diagonal. As the original two arrows are at right angles, the parallelogram becomes a rectangle.

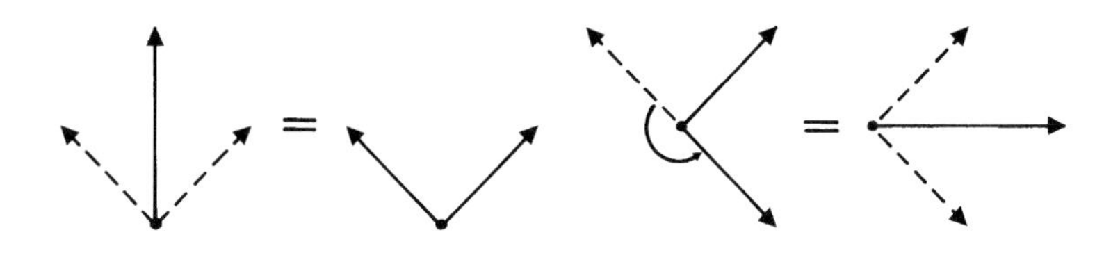

Figure 2.3. Vectors can be used to explain what happened in figure 2.1. Left: vertically polarised light passing into the cellophane film is divided into two equivalent components at right angles to each other and vibrating in the 'privileged directions' of the material. Right: on emerging from the film one wave has been delayed (or retarded) by half a wavelength so that its vector now points the opposite way and the resultant recombined wave is now horizontal, having been effectively rotated through a right angle.

inverted, and when it is recombined with the other arrow, the resultant is at right angles to the original (figure 2.3). As a result all the polarised light is rotated by 90° and passes through the second polariser. In general, the direction of polarisation is rotated by twice as much as the angle between it and the 'special axis' of the film. So turning the jam-pot cover turns the direction of polarisation by twice as much (figure 2.4) and a whole rotation of the film turns the vibrations by two rotations; it ends up unchanged, having been aligned and again crossed with the second polariser four times in the process. This, however, is a special case; if the film is much thinner, giving less than a half-wavelength retardation, then only a proportion of the light is twisted and can pass through the second polariser. A half-wave retardation set at 45° is just enough to twist all the light by 90°, while greater retardation twists an increasing proportion by 180° until a full wavelength retardation leaves all the light vibrating in this direction. This account is therefore somewhat simplified, though not incorrect. A more detailed explanation of what happens with retardations less than or greater than half a wavelength is given in chapter 8.

What is not easily noticed in this demonstration is that not all wavelengths are rotated by the same amount because a given value of retardation can only be exactly half a wave for one particular wavelength. A delay of 287 nm is half a wavelength for yellow light of wavelength 575 nm, but it is 0.64 of the wavelength for blue light of 450 nm and only 0.41 of the wavelength for red light of wavelength 700 nm. (It is sometimes said that this is offset for some materials because the refractive index itself varies with wavelength; but the degree of birefringence, which causes the retardation, is actually greater for shorter wavelengths, thus increasing this disparity. The essential argument, however, is much simpler because a fixed retardation, common to all colours, must necessarily delay each by a different proportion of their wavelength and so affect them differently.) Due to the shape of a simple wave, both the proportions quoted earlier for a half-wave delay give amplitudes that are quite close to those of their respective peaks (figure 2.5) and the resulting rotations are so similar that the differences pass unnoticed. With greater retardations, however, the differences become clear: a 'full-wave' retardation of 575 nm returns the vector for yellow light to its original position (just one wave later) so that it is again blocked by a crossed polariser, whereas red light is turned less and blue light more, so that quite a lot of each gets through and the effect of the mixture is purple.

Brilliant colours are seen when several different films are laid between crossed polarisers. The jam-pot-cover film shown in figure 2.1 is about 20 μm thick and gives a retardation of just about 235 nm—half the wavelength of blue light of 470 nm wavelength; the effect still looks quite uncoloured or 'white' with perhaps a very faint yellowish

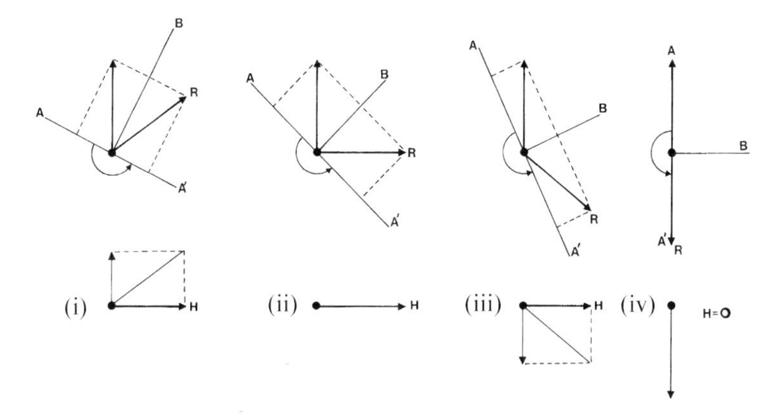

Figure 2.4. Vector diagrams show that as the half-wave retarder film of figure 2.1 is turned, the plane of polarisation is rotated by twice as much. Thick lines show the vectors, thin lines the privileged directions in the retarder and dotted lines are for construction only. In each case the initial vertically polarised light is divided into two components A and B, vibrating at right angles in the 'privileged directions' of the film. Component A is then retarded by half a wave and is effectively inverted to lie along A′. When B and A′ emerge from the film they combine to form the rotated plane of polarisation. Finally, as shown in the corresponding lower diagrams, a horizontal polariser again divides the polarisation into two components, vertical and horizontal, and passes only the horizontal one. The light was originally blocked between crossed polars but in (i) some of it becomes horizontal and is passed; in (ii) the privileged directions reach 45° (as in figure 2.3) and all the light is turned by a right angle; in (iii) the polarisation is turned even further and the result is dimmed, while in (iv) one privileged direction is vertical and has no effect so that the light is once again blocked. A complete rotation of the film 'opens and closes the window' four times.

tinge. But when two such jam-pot-cover films are overlaid, in the correct relative orientation, they give a combined retardation of 470 nm and the emergent light is orange, while three such films give a retardation of 705 nm and the effect is blue (colour plate 1). With thin films that seem to be uncoloured, the individual retardation can easily be assessed by combining several films in this way. A jam-pot cover that is simply folded at random can produce some beautiful colour effects (colour plate 2). Thicker films that give greater retardations and brilliant

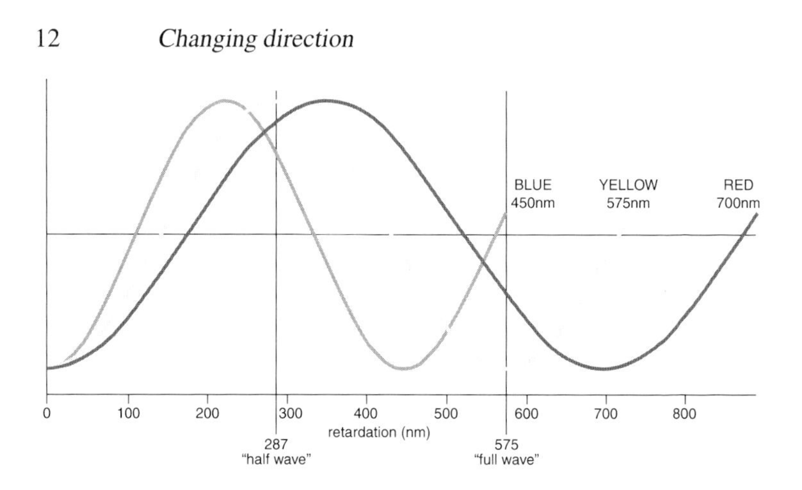

Figure 2.5. The spreading of waves of different wavelengths when subjected to various delays. A 287 nm delay is just half a wave for yellow light, a bit more for blue and a bit less for red. After a 575 nm delay the waves are spreading apart quite significantly. This disparity is actually increased slightly because the delays are generally rather greater for shorter wavelengths.

colours between polars may be obtained from the display wrappers from greetings cards or chocolate boxes (colour plate 3). And of course what is blocked by crossed polarisers will pass through parallel polarisers, so rotating either of the polarisers causes all the colours to change to their complementary colours.

Gradually increasing the thickness of birefringent material, and thus the retardation, produces a sequence of colours as different wavelengths in turn are blocked by the crossed polaroid. The sequence for crossed polarisers runs: black for a film too thin to be effective, paling to white for a 'half-wave' retardation of 287 nm, then yellow, orange, vermilion red and purple for a 'full-wave' retardation of 575 nm. The sequence then continues with a second series: blue, green, yellow, orange, red and a second purple for a 'two-wavelength' retardation of 1150 nm. Further series repeat this latter sequence except that with each repetition the colours become paler. After about the sixth series they are so faded that they are practically indistinguishable. This is because there are so many rotations, and waves of different wavelength become so separated, that no large part of the spectrum is anywhere completely blocked and the result begins to look white again.

When this happens there is actually a series of narrow wavelength bands that are rotated while the bands in between return to their original directions. The latter form dark bands at intervals throughout the spectrum that can actually be seen through a spectroscope. This effect can be easily demonstrated with a compact disc (CD) record that shows rainbow spectra when a bright light is reflected from its back. Shining the light through a 'sandwich' of a retarder that looks white between crossed polaroids (or viewing the CD through such a sandwich) shows dark bands whose spacing depends on the retardation; about 20 jam covers gives three or four dark bands. The bright regions across the spectrum add together to make the material look clear and uncoloured between crossed polarisers, always provided that it is properly orientated; rotating the material still makes it darken every quarter turn. This will be seen again in chapter 3 where thick crystals may be colourless but thin flakes are often highly coloured between crossed polarisers. A really thick crystal, say 1 cm of quartz, gives so many bright and dark bands that their separation may not be possible with the simple CD trick.

The purple colour produced by a full-wave 575 nm retardation is often called the 'sensitive tint' or the 'tint of passage' because a slight increase in retardation makes it look blue while a slight decrease turns it red. If a film of this retardation is superimposed on another material, it can reveal the presence of very slight birefringence (optical retardation) that might otherwise pass unnoticed. The actual values of retardation can be measured by superimposing a graduated 'wedge' of various known retardations and assessing the change in colour. Such retardation wedges are often made from quartz (see chapter 3) and offer values from 'zero' to 2000 nm or more, thus producing three, four or more series or 'orders' of colours. Retardation wedges that can easily be made by hand from gypsum are also described in chapter 3. Colour plate 4 shows an even simpler homemade step-wedge with progressive steps of 55 nm, made by adding successive layers of transparent adhesive tape. Since a retardation in space is equivalent to a delay in time, such a wedge can also be considered as a variable delay line. One step of 55 nm is equal to a delay of 18×10^{-17} s or 180 millionths of a millionth of a millionth of a second! Being able easily to create, measure and control such tiny time intervals using only such simple materials is very satisfying.

The sequences of colours, produced by subtracting different bands in turn from the spectrum in this way, are generally called Newton's colours or interference colours because they also appear in interference effects such as the Newton's rings experiment or in bubble films or when oil is dispersed on water. But in the case of retardation colours in

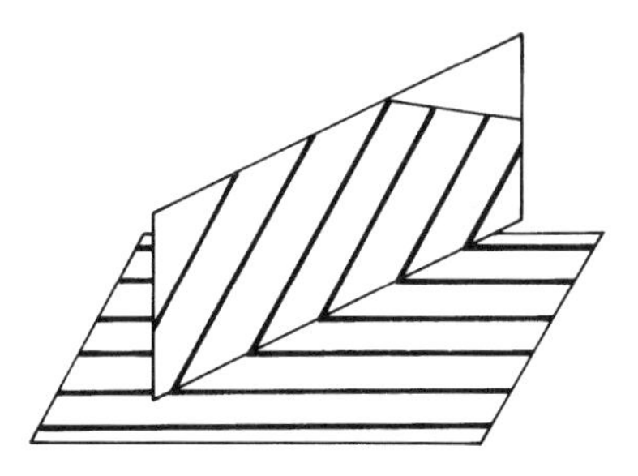

Figure 2.6. A diagram showing how a diagonal mirror (shown upright) can effectively rotate the direction of polarisation (represented by horizontal stripes). This simple geometrical change gives the complementary colours for any given retardation, provided that the mirror and the retarders are all placed between the polarisers.

polarised light no actual interference occurs and attempts to explain the phenomenon by reference to interference are unhelpful and sometimes actually wrong.

> [Two waves polarised at right angles, as in the birefringent film, are unable to interfere at all; when they emerge from the film there is no longer any reason to regard them as separate and they should be resolved into a single resultant; the effect of the second polariser on this resultant can be decided quite simply in the usual way. This avoids some quite elaborate mental and semantic gymnastics. When the retardation is one wavelength so that the two waves once again coincide, the vector returns to its original direction and the resultant wave is blocked by the second, crossed polariser; no interference occurs and the resultant does pass through aligned polarisers.]

Another name sometimes used for this sequence of colours is ‘absorption colours’ and this is quite apt because they are formed when some parts of the spectrum are removed or absorbed, in this case by the second polariser. Here, however, the name retardation colours will be used in order to emphasise the way in which they are produced.

A very striking and instructive effect can be produced when a mirror is introduced between the polarisers, for the colours may be dramatically changed in their own reflections (colour plate 5). This effect is amazing at first sight because no-one ever expects an object to look a completely

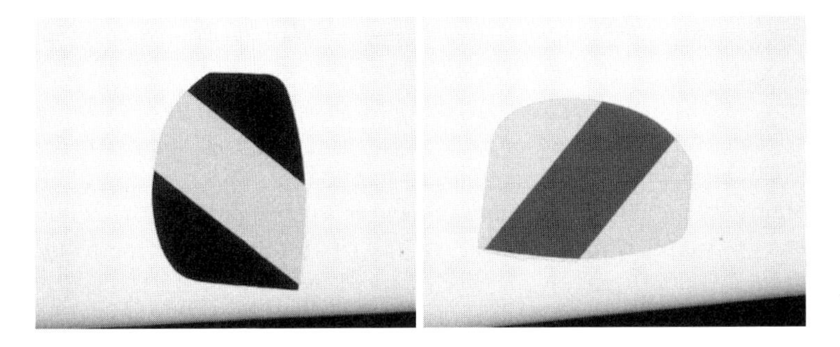

Figure 2.7. One of the most easily improvised polariscopes for detecting polarisation of light, sometimes called Minnaert's design. It can be made by adding a strip of Sellotape diagonally across a piece of polaroid at 45° to its direction of polarisation. The retardation is generally about half a wave (here about 300 nm) and gives a clear contrast in polarised light, except when the polarisation direction is exactly halfway. It is shown in two orientations over a background polariser.

different colour in its own mirror image. The mirror must be held at 45° to the direction of polarisation which, as reflected in the mirror, appears to run away at 90° to the original direction (figure 2.6). The crossed polarisers now appear to be light in the mirror and any colours produced by retardation films are changed into their complementary colours. A more formal demonstration of this is seen with a graded retardation wedge (taken from colour plate 4) and its reflection, as shown in colour plate 6. Of course the colours are not changed if the mirror is held *after* the second polariser, or indeed if the mirror is held parallel to or at right angles to the polarisation direction. Even more mystifying at first sight is the fact that a surface-silvered mirror or a polished metal reflector shows different colours from those shown in a standard back-silvered glass mirror. This phenomenon will be explained in chapter 8.

Retardation colours can be exploited in some interesting ways. In chapter 1 two simple polariscopes were described. Instead of producing a brightness contrast by two polarisers side by side, 'Minneart's polariscope' achieves the same result by a single polariser with a diagonal strip of half-wave retarder film. An easily improvised example is a strip of sellotape placed at 45° across a single piece of polaroid (figure 2.7). The tape, on the side facing the light source, forms a retarder film with a retardation of around 300 nm, acting almost as a half-wave plate and thus producing a strong contrast if the light is polarised. But a visual contrast of two colours is often thought to be more sensitive than a contrast of 'grey' intensities. So another alternative is to use two polarisers orientated at right angles and to cover them both (on the far side) with a retarder film of say 650 nm. Then polarised light will produce a blue colour alongside the complementary yellow (colour plate 7). Simply reversing the device makes the retarder film ineffective so the colours disappear and are replaced by grey contrasts (as seen earlier in figure 1.4). The user can easily compare each method and choose between them. Other colour pairs, say green and red, may be preferred and can be obtained by using different thickness of retarder film.

A quite magical result is obtained when polarisation colours are used in a kaleidoscope. Three mirrors fixed at 60° in the normal way produce a repeated pattern with sixfold symmetry. But instead of using coloured materials to produce the initial image, pieces of clear cellophane of random shape and thickness are jumbled together. Two polaroids, one on each side of the 'specimen chamber', then produce a variety of polarisation colours. When an attractive pattern is seen, rotating one polaroid changes all the colours without altering the pattern. Any gaps between the 'coloured' pieces simply change between light and dark, but if another retarder film is stretched across the whole chamber, these background holes themselves become coloured. Rotation of this film independently of the other elements modulates all the colours in the image, not just the background. A virtually infinite variety of images and colours can be obtained simply by rotating the appropriate supporting collars (colour plate 8).

I once imagined this was an original invention but then discovered that it had been patented in Beijing in 1985. The patent is probably invalid, however, because Sir David Brewster, the inventor of the kaleidoscope, described the method himself in 1858! His book on the kaleidoscope was published in 1819 and the second edition 39 years later had an additional chapter describing just how to use sheets of herapathite and/or a Nicol prism as polarisers and pieces of mica, selenite or other crystals as retarders (all described in chapter 3). He would surely have welcomed a gift of polaroids and cellophane films from the 20th century! Both his and the Beijing instruments placed the second or 'analyser' polariser at the eyepiece so that it can be small and consist of a Nicol prism, say. But this alters some of the colours that are seen after multiple reflection as explained earlier. It is better to place both polaroids in front

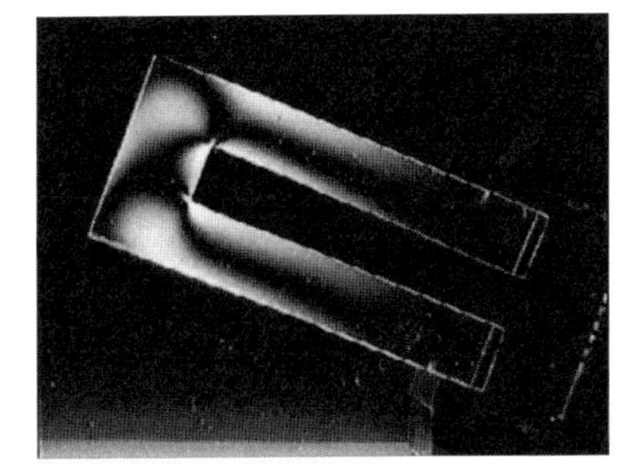

Figure 2.8. A simple U-shaped piece of perspex is normally invisible between crossed polarisers but when the arms are squeezed together gently, the internal strains so produced are clearly revealed due to their birefringence.

of the mirrors although this needs both of them to be as large as the specimen cell itself.

Some polymers such as polymethyl methacrylate (Perspex, Plexiglas etc) do not show birefringence in normal manufactured sheets. But if mechanical stresses are applied, then the internal strains in the material become birefringent and these areas can be seen as light–dark or coloured fringes if viewed between crossed polarisers (figure 2.8). This effect forms the basis for an industrial technology called photoelastic stress analysis. Any engineering component, from a simple lever or a gear wheel to a railway bridge or a cathedral arch, is first modelled in polymer resin such as methacrylate or epoxy. Then stresses are applied to simulate the loads to be expected in real situations and the distribution of internal strains can be analysed in polarised light (figure 2.9). This allows the design engineers to add strength where necessary and save material where possible. Two-dimensional or three-dimensional examinations can be made. In one variant of this technique, some actual components (of steel, say) are coated with a layer of resin and the surface strains are then viewed by reflected light. While in some ways more realistic, this method cannot show internal strains within the material.

Many common objects are made from polymer resins by heat forming or other moulding techniques. In these cases the strains imposed during shaping are retained or 'frozen in' and are easily revealed by viewing between crossed polarisers. Examples abound in any domestic environment and some examples are shown in colour plates 9 and 10.

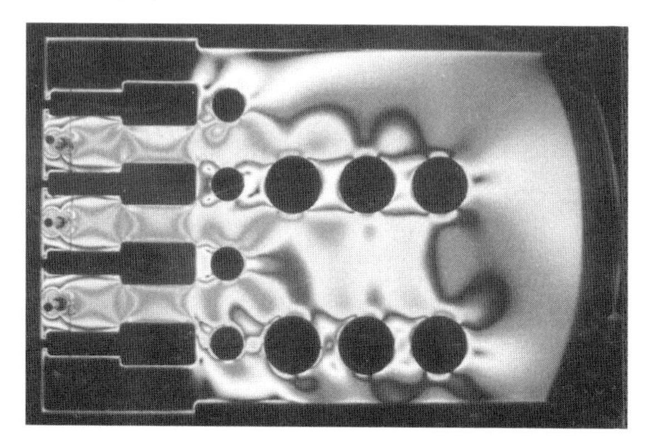

Figure 2.9. An epoxy resin model of part of a large electrical generator viewed in a professional polariscope. The coloured fringes show the strains induced by simulated centrifugal force. (By courtesy of Ken Sharples, Sharples Stress Engineering Ltd, Preston.)

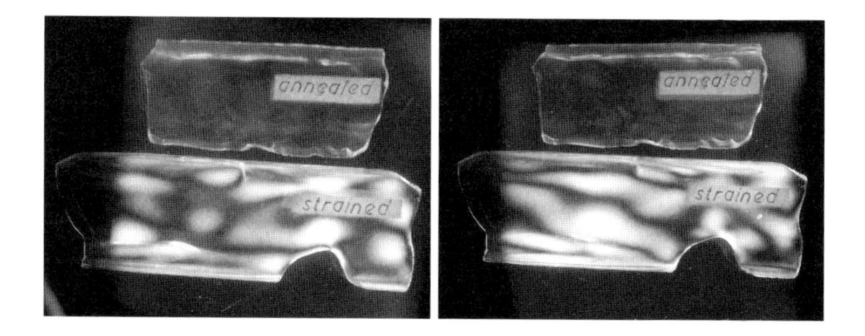

Figure 2.10. Two pieces of worked glass viewed between crossed polarisers. One was allowed to cool immediately and its internal strains show as photelastic fringes; the other was kept overnight in an annealing oven at 565 °C (not quite hot enough to soften the glass) which allowed the strains to dissipate, as shown by the absence of fringes. (Made and kindly loaned by John Cowley, Glass Workshop, Queen Mary, University of London.)

Birefringence also occurs when glass is strained and becomes permanent if the glass is cooled too rapidly after being worked. Such strains make for fragility, so glassblowers often examine their finished work between crossed polarisers and put it in annealing ovens until the strains are relieved. In the example shown in figure 2.10, one specimen was left overnight in an oven at 565 °C, which eliminated all the strains that are still evident years later in the other piece, which had been cooled rather quickly.

Some car windscreens show darkened or coloured patterns when seen through polaroid sunglasses. These screens have been toughened by heat treatment followed by deliberately rapid cooling; the resultant permanent strains ensure that under impact the glass shatters into relatively harmless small granules rather than breaking into sharp shards. The strained regions, however, are birefringent and show up under a variety of circumstances if the driver wears polaroid sunglasses: for example when the incident light is polarised by reflection, say by a wet road (see chapter 7) or comes from the blue sky (see chapter 6). Even light that is not polarised will be partly reflected from the glass and this has a polarising action (see chapter 7), causing the transmitted light to be partly polarised. The patterns may even be seen without polaroid glasses if the windscreen is itself seen by reflection in another window or in the car's paintwork. Many windscreens are strengthened by being laminated instead of being heat toughened and do not show these effects on polarised light. Laminated screens are therefore preferable if the driver likes to wear polaroid sunglasses.

An extreme example of stressed glass is shown by Prince Rupert's drops, so named because they were demonstrated to Charles II in 1661 by Prince Rupert of Bavaria. They consist of molten glass, about 1 cm in diameter, that has been dropped into cold water and so cooled very rapidly. Glass shrinks as it solidifies, so after the outer part of each drop has hardened very quickly, the inner parts cannot shrink as they should and a central space, assumed to be a vacuum, is left. The internal strains are so high that coloured polarisation fringes are very close together (colour plate 11). Although the heads of these glass drops are extremely robust, a slight scratch on the long 'tail' causes the whole object to disintegrate explosively into tiny fragments. They should therefore be treated with great care.

Chapter 3

Crystals

Crystals act on light in some fascinating ways and show many important influences on polarisation. Indeed the early studies of polarised light depended entirely on crystals and they have continued to be of fundamental significance. Crystals can affect light in several different ways and the result is often quite complex, although the basis is quite straightforward.

Crystals consist of a three-dimensional lattice of atoms or ions, all held together with extreme regularity. For instance common salt, sodium chloride, has equal numbers of sodium and chlorine atoms in a perfect cubic arrangement (figure 3.1). A crystal of pure sodium chloride is itself cubic and is formed of a single lattice of such cubic cells, each with an edge length of 0.562 nm. This spacing is fairly typical of crystal lattices which are all around one-thousandth of the wavelength of visible light. Since the sodium chloride crystal lattice is exactly the same in each direction, light travels through it at the same speed in each direction—the crystal is said to be isotropic and it has only one refractive index or speed of light. But other crystal lattices do not have the same structure in different directions: light travels each way at different speeds, so they are said to be anisotropic and birefringent, having two main values of refractive index, a maximum and a minimum. Just as with birefringent polymer films (chapter 2), an anisotropic crystal divides polarised light into two components vibrating at right angles and with different velocities of propagation. One component has the same velocity in all directions but the other has a velocity that varies with direction, either greater or less than the other, depending on the crystal. But, in essence, the long rows of very regularly arranged atoms in such a

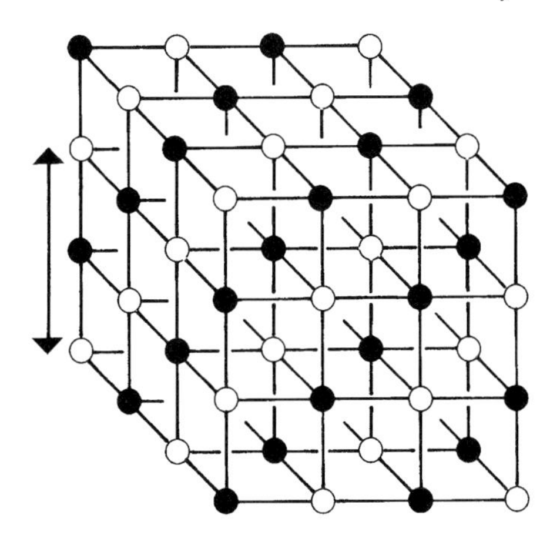

Figure 3.1. The lattice structure of a simple cubic crystal of sodium chloride. Positive sodium ions (charged atoms—dark) and negative chloride ions (charged chlorine atoms—pale) are held by electrical forces to form a regular cubical pattern with a repeat distance, as shown by the arrow, of 0.562 nm (one-thousandth of the wavelength of yellow–green light). On an enormously greater scale, such a lattice forms a crystal that is itself cubic.

crystal can act on light just like the long, parallel molecules of polymers.

Ice is crystalline in structure and a spectacular demonstration is produced by rapidly freezing a shallow dish of water (by pouring liquid nitrogen onto it) between crossed polaroids on an overhead projector. Initially the liquid water is isotropic but as the ice crystals grow, they are birefringent and show up in brilliant colours, each according to its own orientation until they all meet within the solid mass. The example shown in colour plate 12 was frozen more slowly in a freezer compartment. Another nice example is salol (phenyl salicylate) which is very strongly birefringent. Crystals can easily be melted (at 43°) on a glass plate and another warm plate is then pressed onto the melt. As the sandwich cools, the salol recrystallises in a very thin layer that shows splendid colours between crossed polarisers (colour plate 13). As explained in chapter 2, thicker birefringent crystals show no retardation colours because many wavelengths across the spectrum are rotated and all the wavelengths in between are not. So the crystal looks clear between polarisers, whether crossed or parallel, provided that it is properly orientated (figure 3.2). Rotating thick crystals extinguishes the light every 90°, as happens with

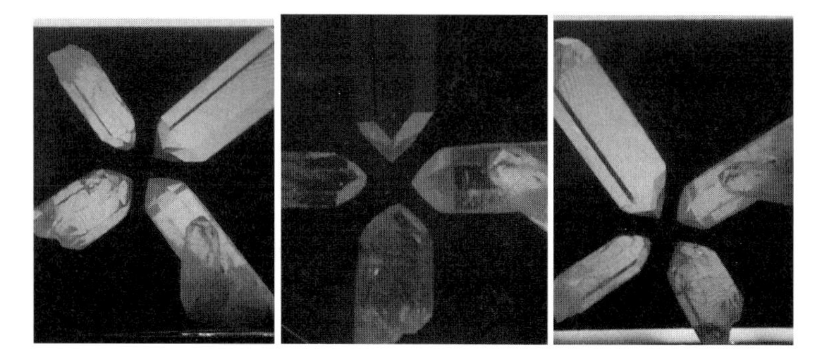

Figure 3.2. Quite large quartz crystals between crossed polaroids. As the crystals are turned they become transparent four times for each rotation as they turn the direction of polarisation; at the intermediate points they can only be seen by reflected background light. No colours are seen, however, unless the crystals are very small (the same material is seen on a microscopic scale in colour plate 16).

a half-wave plate, but the explanation is rather different.

The term 'dichroism' originally referred to crystals that simply looked to be different colours (or clear) when viewed along different axes; indeed the word literally means 'two coloured'. But the effect is often much clearer when different directions of polarisation are used in viewing the crystals. In some cases. one component of the light is absorbed (the crystal is more or less opaque to it) whereas it may be quite transparent to the other component. A good example of a crystal of this kind is tourmaline. As shown in figure 3.3, two thin pieces of tourmaline act just as polaroid film: they are fairly transparent to green light (the colour varies between specimens) but when they are crossed, the combination is quite opaque. Slices of unflawed tourmaline crystals were often used as polarising components in optical instruments (figure 3.11) as they were cheaper than Nicol prisms (described later) although they were generally of an inferior optical quality and their self-colours were sometimes undesirable. Another well-known example of a naturally occurring dichroic crystal is epidote. The artificial crystals of herapathite (iodo-quinine sulphate) were described in chapter 1 as they were a component of early kinds of polaroid.

In other crystals, one component may have only some of its wavelengths absorbed so that it emerges coloured, while the other

Figure 3.3. Tourmaline is a dichroic crystal: left, slices of green tourmaline one parallel to and one crossed with a background polaroid; centre, the same two pieces, both turned by 90°; right, the same two slices of tourmaline crossed with each other in normal light, with no other polariser.

component is clear. An example is sapphire which is deep blue for one direction of polarisation and clear for the other; since the eye cannot distinguish the different polarisations, they are seen mixed together and the effect is a paler blue. Obviously, any polariser allows one to distinguish immediately between a real sapphire and an isotropic crystal or blue glass. The same relation applies to the red colour of rubies. Alternatively, in some crystals the two components may both be coloured, but of different hues. A good example is copper acetate which is a bluish-green in colour, but when viewed through a polariser the colour changes from deep, royal blue to clear light green as the polariser is turned (this difference is shown in colour plate 14).

Even more variety is added by crystals that are different along three axes rather than two—a property called 'trichroism' or 'pleochroism' (literally 'more colours'), associated with three significant values of refractive index in different directions. These crystals are sometimes different in colour when simply viewed along each axis by unpolarised light (tourmaline sometimes shows this property). But again such crystals may also show quite different responses to polarised light in the different directions. Each self-colour may turn out to have two components under a rotated polariser or the transmitted light may be absorbed for one direction of polarisation. These effects may be different for each axis of the crystal. In other words each axis may be dichroic in a different way from another axis.

Even when there is no dichroism, most crystals (all except cubic ones) show some degree of birefringence and thus affect polarised light. One of the most birefringent of natural crystals is calcite or Iceland spar (calcium carbonate) whose birefringence was described by the Danish scientist Erasmus Bartolinus in 1670, in what seems to be the first ever observation of an effect due to the polarisation of light. This material is one of the major constituents of the earth's crust, usually in microcrystalline form in marble, limestone, chalk or coral but sometimes as large, clear, rhombic crystals which are found in Iceland ('Iceland spar') and Mexico. The degree of birefringence is expressed by the difference between the two refractive indices, which for calcite are 1.486 and 1.658, giving a large difference of 0.172. Sodium nitrate ('Chile saltpetre') has an even larger birefringence of 0.251 but it is much less convenient to experiment with as it readily dissolves in water and so can easily be disfigured by handling (for this reason it only occurs naturally in very dry conditions as in Chilean deserts). In both these cases the two refractive indices are so different that the two refracted rays can be seen to diverge very markedly. A calcite crystal placed over a dot or other mark on a piece of paper shows two images (figure 3.4) and a sodium nitrate crystal does the same (figure 3.5). If the crystal is rotated, one image stays still while the other one moves in a circle around it. It was this observation, first made by Bartolinus in 1670, that eventually led to the discovery of polarisation and, in turn, contributed greatly to our basic understanding of the nature of light itself. We now know that the 'ordinary ray', which gives the stationary image, has a lower velocity within the crystal than the 'extraordinary ray' that gives the moving image. In some crystals it is the other way round—the ordinary ray is faster.

Viewing the double image through a polariser shows that the two images are polarised at right angles to each other because turning the polariser brightens one image and extinguishes the other in turn. If the dot on the paper is replaced by a small hole in a black card, a polariser can be placed over the hole itself and again the two images seen through the crystal can be extinguished in turn, as shown in figure 3.5 with a crystal of sodium nitrate. A calcite crystal combined with a lens to view the double image of the hole (figure 3.6) makes a kind of polariscope called a dichroscope that is used by jewellers. Any dichroic

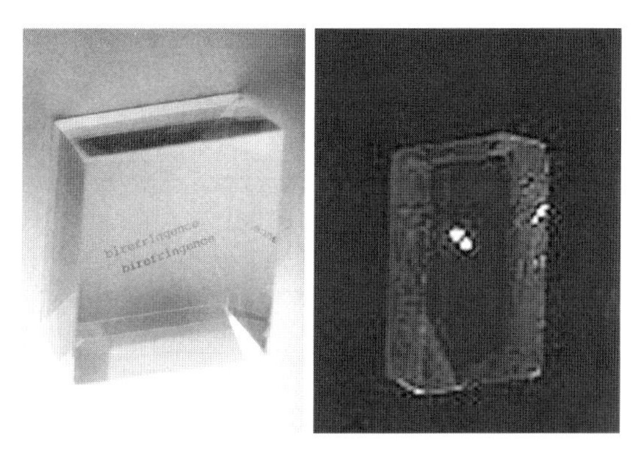

Figure 3.4. Left: a calcite crystal over a single typed word on a sheet of paper, showing a clear double image. This spectacularly large, clear crystal belongs to the Royal Institution, whose help is gratefully acknowledged. Right: a smaller calcite crystal over a small hole in a black card. In both cases, rotating a polariser over or under the crystal would extinguish each image in turn, showing that they are polarised at right angles to each other.

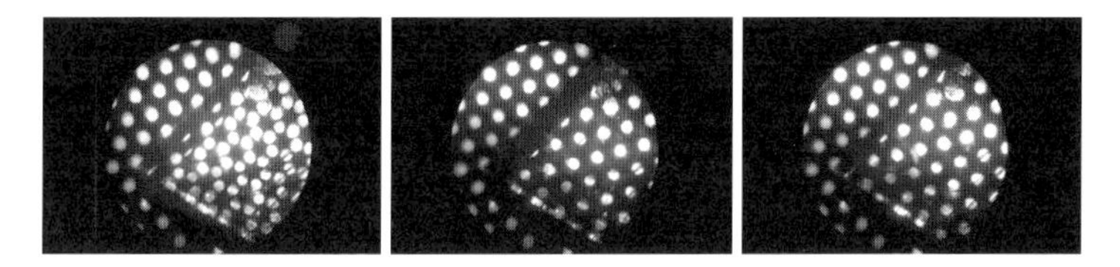

Figure 3.5. Left: a small sodium nitrate crystal over a regular array of holes in a black background. Each hole seen through the crystal creates a double image. Centre: the same seen through a sheet of polaroid that suppresses half the images. Right: the same again but with the polaroid turned by 90° to suppress the other set of images instead. Clearly the two sets of images are polarised at right angles to each other.

material placed in front of the hole can give different effects side by side in each image; thus a sapphire gives one blue and one clear image simultaneously, while copper acetate gives one blue image alongside a green image (colour plate 14). Materials with only one refractive index, such as glass, cannot produce such differences in colour between the two images of a single hole.

The impression is often given that calcite crystals also produce a double image of distant objects, but simply looking through the crystal

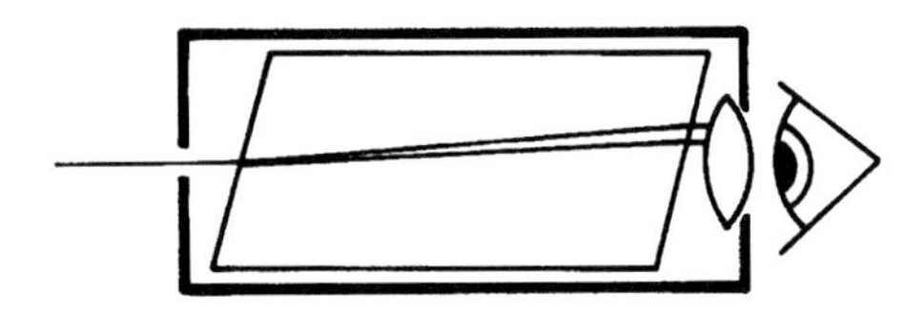

Figure 3.6. A calcite crystal can be made into a simple dichroscope that can be used to reveal dichroism in other specimens. A single small hole, seen through a calcite crystal and a small lens, produces two images that are polarised at right angles to each other. An improvised instrument was conveniently housed in a black plastic film can, and to get a neater edge the hole was drilled in a sheet of metal that was then fixed over a larger hole on the end of the case.

does not work. If one watches through a calcite crystal as a dot or hole is steadily moved further away, the spacing between the images appears to diminish with distance just as if they really are a double structure. The reason is that the two polarised rays diverge within the crystal but when they emerge again they become parallel, although separated by a little over 1 mm for each 10 mm of crystal thickness. So a spacing of, say, 2 mm from a fairly large crystal is easily seen when close to the crystal but it becomes insignificant at a distance of more than 1 m and distant objects just do not look double. But if one looks down into the crystal so that the image is seen after it has been reflected in the intermediate face (figure 3.7), then two images can be seen, each polarised at right angles to the other and at 45° to the plane of the incident and reflected rays within the crystal. This is because the two emerging rays end up diverging by about 20° and produce two well-separated and oppositely polarised images. Figure 3.8 shows how this can be demonstrated and figure 3.9 shows the result.

But this technique is rather inconvenient—for one thing the two images overlap extensively and for another the field of view is very restricted because either the rays entering the crystal or the emerging rays (or both) are close to glancing angles to the respective crystal faces. To make a more practical use of the birefringence of calcite, special 'double image prisms' have been invented. In one design, a 60° prism is cut from a calcite crystal and, instead of producing a spectrum of rainbow colours from each point of the image, it produces two spectra in different directions, each polarised at right angles to the other. A glass prism can then recombine the colours of each spectrum to produce two

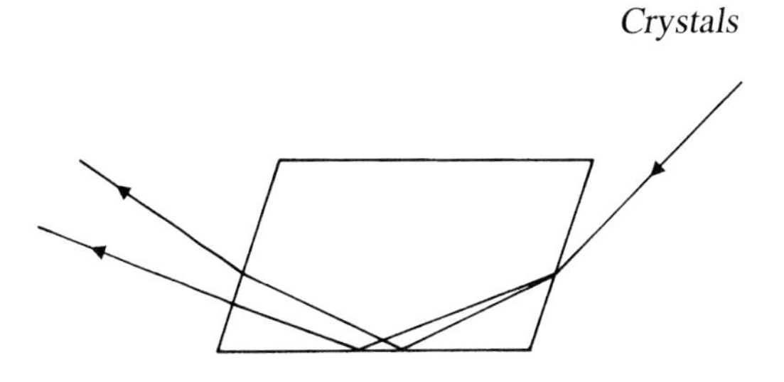

Figure 3.7. A ray of light entering a calcite crystal so that it is reflected at the next face produces a double image because birefringence results in two divergent beams and after refraction they continue to diverge. By looking down into the third face, one can see the two images and simple tests with a piece of polaroid show they are polarised at right angles to each other and at 45° to the plane of the diagram (see figure 3.9). As with most figures in this chapter, the angles are not all depicted accurately but have been altered to help clarity.

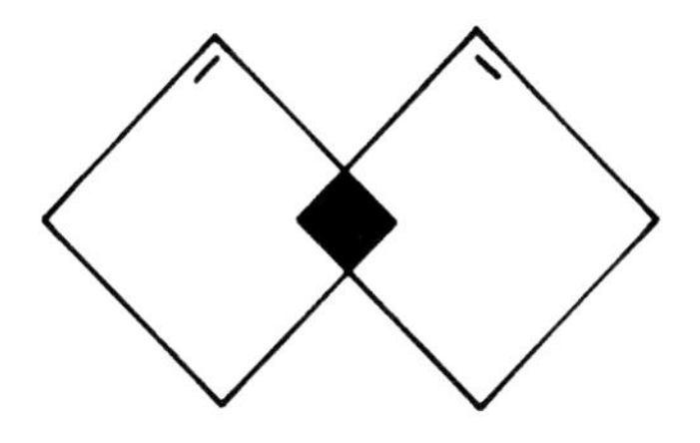

Figure 3.8. The setup used to photograph the polarised images produced by a calcite crystal as explained in figure 3.7. A light box, consisting of a backlit translucent screen, had two square polaroids mounted on it. Their oblique directions of polarisation were set at right angles as shown by marks in the top corners and by the small central area of overlap.

virtually uncoloured images, polarised at right angles to each other and well separated in space.

In 1828 William Nicol realised that one of the divergent beams within an ordinary calcite crystal could be eliminated by using the principle of total internal reflection. He cut across a crystal at a carefully calculated angle and cemented the two halves together again with Canada balsam (figure 3.10). If the angle of the cut is just right, one beam is reflected out through the side of the crystal while the other one proceeds, giving a complete separation of the two polarised components. This design was capable of several modifications (notably one by Sylvanus P Thompson) which together formed the best polarising

Figure 3.9. A photograph taken through a calcite crystal as in figure 3.7 of the light source shown in figure 3.8. One polaroid appears darkened and the other clear, showing that the whole image had become polarised at 45°. The other image, which emerges from the crystal at a very different angle, is exactly the same except that the condition of the two polaroids is reversed, showing that it is polarised at right angles to the above image. (Reflection, discussed in chapter 7, does not affect the polarisations here because both directions are oblique to the reflecting surface.)

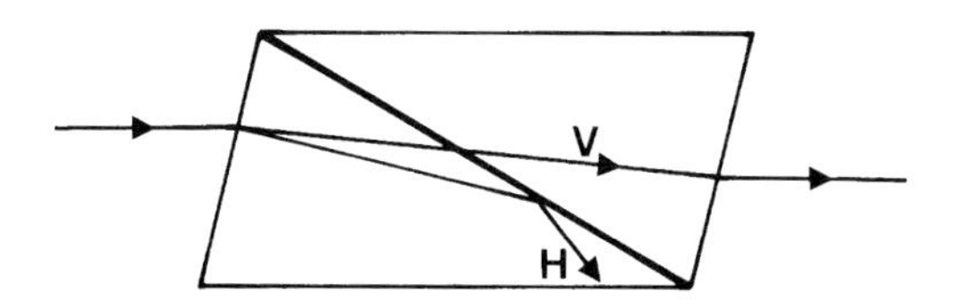

Figure 3.10. A Nicol prism polariser is made by cutting a calcite crystal at the proper angle and cementing the two pieces together again with Canada balsam. Light entering the crystal is divided by the strong birefringence into two divergent rays that are polarised at right angles. The horizontally polarised beam (H) is totally reflected at the sloping interface while the vertically polarised beam (V) continues through the crystal and emerges as 100% polarised light.

components available until polaroid became available in the 1930s. A typical Nicol prism, mounted in brass for use in an optical instrument, is shown in figure 3.11 together with a mounted tourmaline which was a cheaper option. Even today Nicol prisms are often used when the finest optical quality is needed. But they do have drawbacks: very clear crystals, free of flaws, must be used and these are seldom large so that the aperture is restricted; they only work for light beams that lie close to the long axis of the prism, so they have a narrow acceptance angle, and they

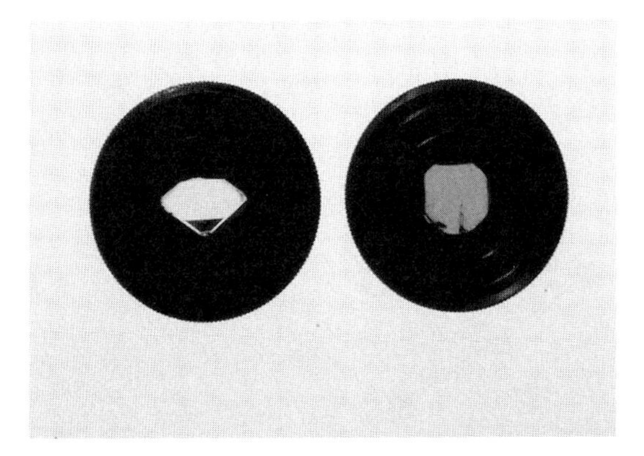

Figure 3.11. A Nicol prism (left) and a tourmaline crystal (right) mounted for use as optical devices, perhaps in a polarising microscope. Both are seen aligned with a background of polaroid. When turned at right angles, both devices become extremely dark, almost opaque.

must be very carefully crafted to make the crystals into actual prisms. The combination of scarce natural material and fine workmanship makes large Nicol prisms very expensive. The double refraction of calcite has also been exploited in other designs for polarising prisms but sheet polaroid has superseded them all for most purposes.

The reason for the very high degree of birefringence in calcite is suggested by the lattice structure of the crystal. The formula for calcium carbonate is $CaCO_3$ and within the crystal the calcium ions and the carbonate (CO_3) ions form separate alternating layers. Each carbon atom is surrounded by its three oxygen atoms in a common plane with other carbonate ions (figure 3.12). These layers of negatively charged ions are linked by electrostatic forces to the intervening layers of positively charged calcium ions and such electrical fields interact with light waves which are themselves electromagnetic in nature. Finally, these atomic lattice planes are at an angle to all the faces of the rhomboidal crystal, so that any light passing between two opposite faces must pass at a slant across millions of sloping planes of interactive charged layers. It is not necessary to understand the details of the interaction in order to expect that components of the light wave polarised in a direction normal to the lattice planes will have a different (actually greater) velocity than those polarised parallel to the lattice planes: i.e. the crystal will be anisotropic.

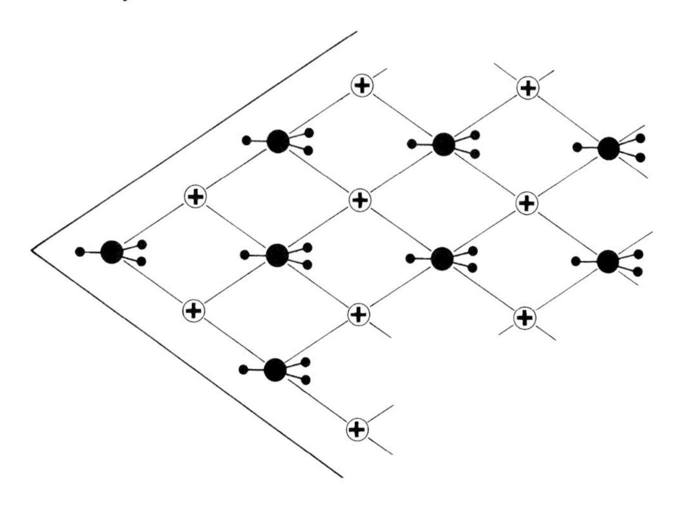

Figure 3.12. The lattice structure of calcite ($CaCO_3$). The horizontal planes of positive calcium ions (shown clear) are interleaved with planes of flat negative carbonate ions (shown as dark carbon atoms each attached to three oxygen atoms). They are bonded together by electrical forces, with each calcium attracted to its nearest three carbonates in the plane above and to three in the plane below (only two can easily be shown in a two-dimensional diagram). These electrical fields interact with the electromagnetic waves of light passing through them at an angle; this happens for light crossing between any opposite faces of the normal rhomboidal crystal, one edge of which is indicated by the straight lines, and accounts for the strong double refraction.

The same explanation applies to sodium nitrate ($NaNO_3$) where sloping nitrate planes are similarly interleaved with planes of sodium atoms.

Another natural crystal that is much used in optical devices is quartz or silicon dioxide. This is even more common in the earth's crust than calcite, since silicon and oxygen are the two most abundant elements and they are largely combined as the dioxide to form a constituent of many rocks, sand and sometimes large, clear 'rock crystals' or coloured amethysts. The birefringence of quartz is only about one-nineteenth that of calcite (0.009 instead of 0.172). But when an application calls for small retardations such as half-wave or quarter-wave plates, the appropriate slices of calcite would be too thin and fragile. Quartz is a much stronger material and the equivalent quartz sheet is also much thicker which makes it stronger and easier to work to the

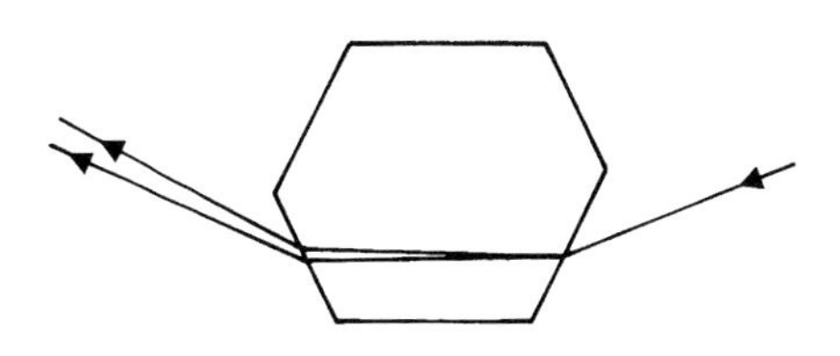

Figure 3.13. Using a columnar quartz crystal as a prism. A beam of light entering the crystal at the right angle to leave through the next-but-one face is divided into two slightly divergent beams that are polarised, one parallel to the axis of the crystal and the other at right angles. The divergence, here shown exaggerated for clarity, is so small that images viewed this way show considerable overlap as well as coloured fringes.

required accuracy. 'Quartz plates' are therefore common accessories for polarising microscopes (see later). Graded retardations are produced by a 'quartz wedge', a much superior component to the simple step-wedge described in chapter 2. Other retarder components are sometimes made of gypsum ('selenite plates' of calcium sulphate: birefringence about 0.01) or of mica (birefringence about 0.036), both of which are easily cleaved to the right thickness. An improvised retardation wedge made of gypsum is described later.

The trick of looking through a calcite crystal at an image reflected internally at one face can also be done with quartz. But because quartz is hexagonal, one can also look at an image that has been refracted from one face of a quartz crystal so that it emerges through the next-but-one face (figure 3.13). In effect the column of the crystal acts as a prism but gives two overlapping images. The divergence of the emerging rays is here very small and most images overlap so much that they look blurred rather than simply double, and of course there are also strong colour fringes. Nevertheless if one looks at a narrow slit or a linear light source (figure 3.14), there are two separate images very close together and polarised at right angles (figure 3.15).

The birefringence of a quartz crystal can also be demonstrated by using it as a prism to throw a spectrum onto a screen. Draw the blinds of a sunny window, leaving a narrow gap, and hold the crystal there so that the sun shines on it. A patch of light, coloured by a rainbow-like spectrum, will be thrown somewhere on the walls or ceiling. But the appearance of this coloured patch will be rather unfamiliar. Instead of a single patch, as with a glass prism, there are two overlapping patches

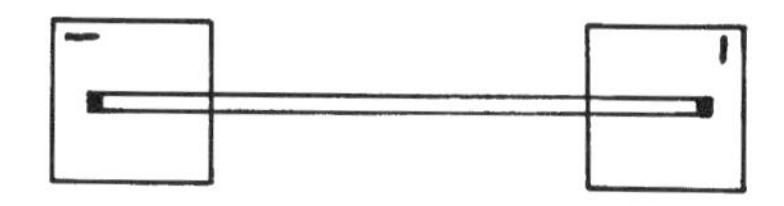

Figure 3.14. The setup used to photograph the double image that can be seen when a quartz crystal is used as a prism, as shown in figure 3.13. In order to avoid overlap between the images, a narrow source is necessary. Here a rather distant tubular fluorescent lamp was covered with a polaroid over each end, one with its polarisation direction vertical and the other horizontal as shown by the marks in the corners.

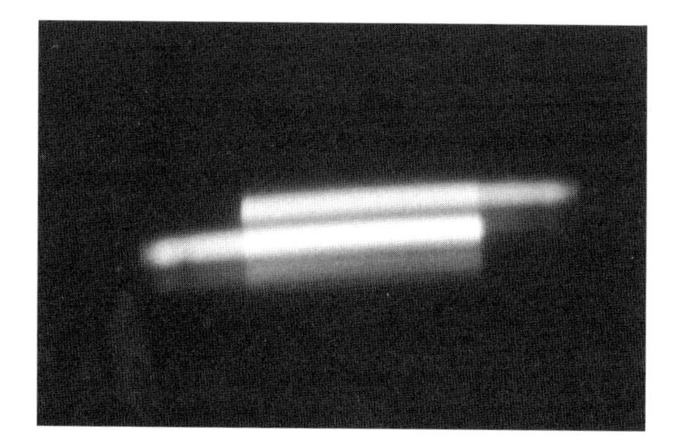

Figure 3.15. The double image of the lamp shown in figure 3.14 photographed through a large quartz crystal kindly lent by Stuart Adams. The apparent lateral displacement between the images is due to the fact that each is polarised at right angles, one vertically and the other horizontally. So the polaroid at the left is opaque in one image and the polaroid at the right is opaque in the other.

just over 1° apart (figure 3.16)—about 2 cm at a distance of 1 m. By turning a piece of polaroid in the light path, each spectral patch can be extinguished in turn, showing the normal sequence of colours and also proving that they are polarised at right angles to each other. When the same experiment is done with a calcite crystal, the two patches are separated by about 20° (i.e. 40 cm at 1 m) due to the much greater birefringence. They are so far apart that one of them might be taken for a stray reflection within the crystal, and it is only by using polaroid to extinguish them alternately that they can be clearly related to each other.

Quartz also shows, much more than other natural crystals, another

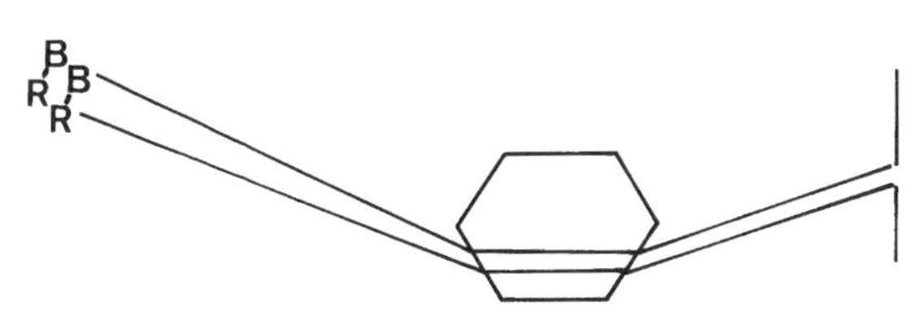

Figure 3.16. Sunlight can be refracted by normal-shaped crystals to produce rainbow spectra. A 'pencil' crystal of quartz placed in sunlight, passing through a slit on the right, makes two spectra that just overlap as they diverge by just over one degree. They are polarised at right angles to each other, and a polariser rotated anywhere in the light path can remove either of the spectra and so clarify the sequence of colours (B—blue to R—red). A calcite crystal gives a much larger separation of around 21° due to its much higher birefringence.

property, called optical activity. Crystals do not show birefringence for light passing along their optical axis—the line normal to their symmetrical lattice planes. In cubic crystals all three directions are optical axes so there is no birefringence at all, but anisotropic crystals have one or two axes for which this applies (when there are two they are divergent, not at right angles). Nevertheless, when polarised light passes along the optical axis of quartz it is steadily rotated, depending on the length of its path through the crystal. The mean rotation is about 21° per millimetre of crystal but it depends rather strongly on wavelength, so that deep blue light is turned three times further than red (50° and 16° respectively). When such an optically active crystal is viewed along its optical axis and between crossed polarisers, therefore, it gives no extinction but shows colours that change as one polariser is rotated. Unlike the 'twisting' effect of birefringence (see chapter 2), light subject to optical activity remains linearly polarised at every stage so that the effect really is a simple rotation of the polarisation direction alone. Some quartz crystals rotate it clockwise and some anticlockwise. A few other kinds of crystal also show optical activity but not usually as strongly as quartz (an exception is cinnabar, mercury sulphide, which is about 20 times more effective than quartz). Some cubic crystals, which are not birefringent at all, show optical activity along all three of their axes and sodium chlorate is a good example, giving about 3° per millimetre. The explanation of optical activity will be presented at the end of chapter 5 in which the reason for the phenomenon is discussed in some detail.

Geologists exploit the fact that most crystals have distinctive effects on polarised light which can be measured fairly easily and are catalogued in reference tables. Most igneous rocks consist of a mixture of very fine crystals of different minerals. If the rock is carefully ground down to a slice about 30 μm (thousandths of a millimetre) thick, then light can shine through the individual constituent crystals (colour plate 15). Examining these rock sections between crossed polarisers under a microscope not only makes a beautiful display of retardation colours but allows each crystal to be characterised and, in conjunction with other physical attributes, to be identified.

Geological microscopes for quantitative work vary in their arrangement according to the preferences of their designers and makers. But basically they all have arrangements to introduce polarisers both below and above the specimen stage and also to include various accessories such as calibrated retarder plates. With just a little ingenuity, any basic microscope can be adapted to produce striking effects and, with care, even quantitative measurements. A now obsolete device that was once used for such work was called a 'selenite stage', referring to its use of gypsum retarder plates. It was placed on the microscope stage and had a slider that could insert a polar of tourmaline and a gypsum or mica retarder directly under the glass slide bearing the specimen. It thus avoided problems with polarised reflection from the mirror (chapter 7) and birefringent stresses in the substage condenser lenses (chapter 2). With the advent of sheet polaroid instead of bulky Nicol prisms or coloured tourmalines, such a special slide carrier is quite simple to improvise and makes observations less fiddly, although all the effects may also be easily seen without it. A proper geological microscope also has a rotating specimen holder with a built-in protractor but this is only necessary for making actual measurements of optical constants. The second polaroid can be fitted to a cap on top of the eyepiece where it is easily rotated. The photographs in colour plates 15, 16, 19, 20, 21 and figure 3.17 were all made with such a home-made set-up. It is not even necessary to have thin rock sections, which are difficult to prepare and quite expensive to buy, because small clear crystalline fragments such as silver sand (rough quartz) give splendid effects without any preparation at all (colour plate 16).

The professional geologist uses three basic accessory retarders. A quarter-wave plate, once commonly made of mica, enhances or diminishes the retardation of an observed crystal and is used to distinguish between the directions of the ordinary and extraordinary rays in a specimen. A full-wave plate, usually of gypsum, was sometimes called the ‘tint plate’ because, between crossed polarisers, its purply red colour (the ‘tint of passage’ or ‘sensitive tint’) is unmistakably reduced to red or increased to blue as a sensitive detector of quite small crystal birefringences that might not be observed directly. Both these accessories can easily be improvised by finding cellophane films with retardations of about 150 nm and 575 nm respectively (chapter 2). Plain films with other retardations are worth trying for their aesthetic effects (colour plate 15) although they have no real practical value. Pieces of mica can be obtained from some capacitors in older radio receivers (colour plate 17) but they are generally rather small and have no real advantages over films. The mounted films shown in colour plate 5 were made for use with a microscope stage and were used in preparing colour plate 15.

The third standard accessory is a calibrated quartz wedge of continuously varying retardation, from near zero to about 1700 nm or even 2400 nm, thus covering all colours up to the third or fourth order. This component is used to make quantitative measurements by overlapping it with a specimen crystal and observing the change of the retardation, as on a scale. It is not easy to improvise a quartz retardation wedge because quartz is very hard and difficult to work—it is harder than a steel knife blade and will scratch glass. Gypsum is very much softer and gives almost the same retardation, so quite respectable optical retardation wedges (colour plate 18) can be made by carefully rubbing thin slivers of gypsum on a fine stone and examining them frequently between crossed polars to check progress. Because gypsum is so soft and fragile, it is difficult to grind the thin edge to give less than about 250 nm retardation but this can be remedied by simply superimposing an appropriate cellophane retarder film over the whole wedge. The film is orientated so that it counteracts the effect of the thin end of the crystal, that is until the end becomes black between crossed polarisers. All other retardation values are reduced by the same amount of course and the sequence of colours ‘slides’ along the wedge. A similar trick is often used even in professionally made quartz wedges. Cruder step-wedges can be made from successive layers of adhesive tape (chapter 2, colour plate 4).

As an alternative to a wedge, very small calibrated changes of retardation can be introduced by inserting a mica plate into the light path and then tilting it to increase its effective thickness. This ‘Senarmont compensator’ can be made extremely sensitive but it cannot fit close to the specimen and so needs to be placed within the tube of the microscope, which therefore requires rather major modification. Another method is to superimpose two different retarders, one of them quite fine such as a quarter-wave plate that can be rotated so as to add to or subtract from the larger retardation. The rotating ring can then be calibrated to give the combined retardation in nanometres. Several more elaborate compensators have been designed and manufactured for professional use.

Geologists also characterise crystals by the interference figures they produce in convergent beams of light—also called conoscopic figures. This requires the microscope to be fitted with simple accessory lenses. Although these figures are due to polarised light, their explanation is outside the scope of this book but can be found in any text of optical mineralogy or crystallography.

Ordered, crystalline structures are found in many natural, organic materials. Structures within animal and plant tissues, as well as products derived from these, show some beautiful retardation colour effects when viewed under a polarising microscope. Examples include hair (colour plate 19), tendon and silk (colour plate 20) from animals and cell walls (and hence kapok and cotton: colour plate 21) from plants. Less surprisingly, perhaps, many cell inclusions are crystalline, including starch grains (figure 3.17) that form the food stores of plant cells. The polarising microscope is therefore a valuable tool for detecting any regularity in molecular structure. But although it is widely used by a variety of specialists, the simple basic techniques deserve to be tried much more generally. They are, after all, very simple to improvise and they can then yield results that are not only beautiful and diagnostic but can easily be made quantitative and highly accurate if necessary.

Crystals of a rather different kind, now in widespread use, are liquid crystals. These are composed of molecules that, between certain temperature limits, spontaneously align themselves in one direction. They are not rigidly bound as in a solid crystal, yet they are not totally disorganised as in a melted liquid. The nature of liquid crystals was first made clear, following their discovery in 1888, by studying them under polarised light. There are several types of liquid crystal but some are birefringent and so twist the direction of polarisation of light, while some are asymmetrical (see chapter 5) and produce circularly polarised light (see chapter 8). The individual molecules may also align themselves with an electric field that then determines the crystal axis, and this is the basis of the liquid crystal displays (LCDs) now used in most pocket calculators, digital watches and miniature TV screens. The details vary

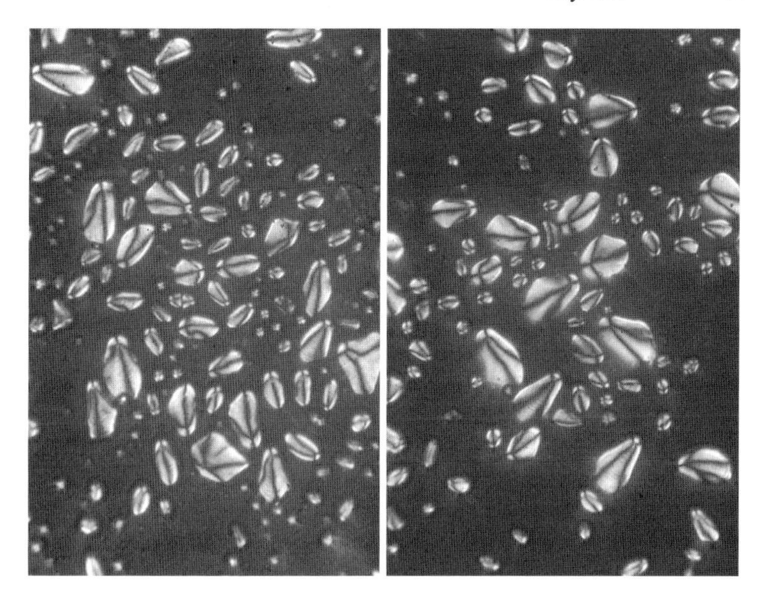

Figure 3.17. Starch grains are the food reserves stored by many plant cells and may ultimately form a staple diet of humans. They have a crystalline structure that is laid down as successive layers around an initial point called the hilum, so that the optical properties radiate outwards in all directions. Between crossed polarisers, therefore, the grains reproduce all the successive views of a rotated crystal: four bright bands alternating with four lines of extinction. In the grains of potato cells, seen here under the microscope, growth is unequal and the grains are oval and excentric, with the hilum towards one end, unlike grains of wheat starch which are spherical with the hilum at the centre.

between different designs but typically a thin cell (about 10 μm thick) of liquid crystal is placed between parallel polaroids. If the optical axis is normal to the polaroids (along the light path) the cell remains transparent but an applied voltage can rotate the molecular crystal axis by 90° to lie across the light path. The cell is then birefringent and acts as a half-wave plate, so the plane of polarisation is rotated 90°. The cell then blocks the light path and looks black.

A set of seven such cells can be used to form all the familiar numerical characters from 0 to 9, while more elaborate arrangements are able to produce alphabetical characters or more complex images. One striking property of LCD displays is that they consume extremely small amounts of power; only 1 watt would be needed to maintain 100 m^2 in the 'switched on' condition. This makes them very attractive for use in pocket calculators for instance, especially those powered by small solar cells. One can easily detect the presence of polarisers in LCD displays by viewing them through another polaroid and rotating it: the screen darkens and lightens as would be expected. But the design of LCD displays is sometimes unfortunate: those used in the instrument panels of cars may not be aligned with polaroid sunglasses worn by the driver so that their apparent brightness is degraded and they can be completely blocked by a slight tilt of the head.

Chapter 4

Fields

Michael Faraday is widely regarded as the greatest experimental scientist ever. He not only showed great skill and ingenuity in the laboratory, he also brought deep insight to bear on the problems he studied. He was able, far more than most, to see what issues were worth studying and to persevere with them even when he met with persistent failure.

From 1820 to 1831, working at the Royal Institution in London, Faraday established the laws of electromagnetic induction: the relationship between magnetism and electricity which produced the electric motor, the dynamo and the transformer. He then took up the subject of electrochemistry in which he established some fundamental relationships between electricity and matter. From all this he came to believe that all physical forces and phenomena must be related, even interchangeable—he would have had great sympathy with modern attempts to find a 'grand unified theory of everything'. In 1845 he wrote:

> I have long held an opinion, almost amounting to conviction... that the various forms under which the forces of matter are made manifest have one common origin; or in other words, are so directly related and mutually dependent... this strong persuasion extended to the powers of light... and (I) have at last succeeded in *magnetizing and electrifying a ray of light*.... (his own italics)

After exhaustive tests over a long period with completely negative results, on 13 September 1845 he tested some samples of glass that he had made some years earlier under a contract from the Royal Society. He found that when light was passed through a piece of very dense

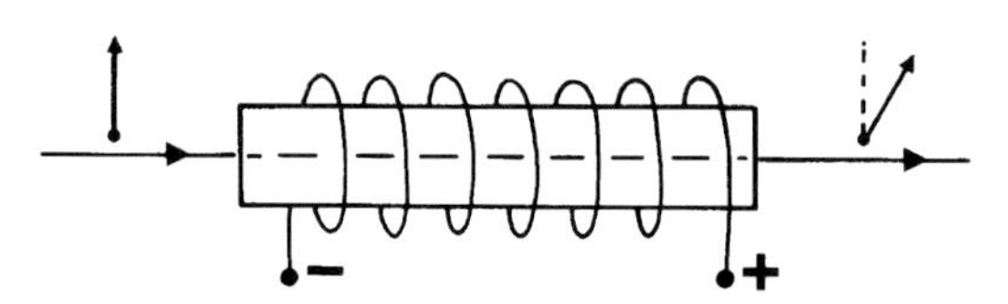

Figure 4.1. Faraday magneto-optical rotation. A rod of very dense glass is placed along the axis of a magnetic field, here represented by the coil of an electromagnet. Polarised light passing along the rod is rotated as indicated by the vector arrows.

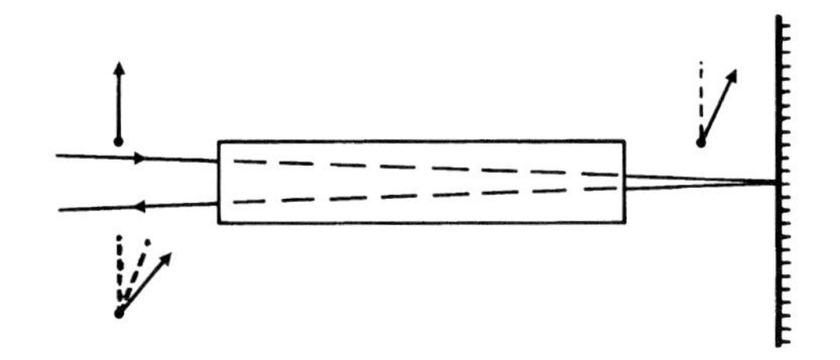

Figure 4.2. If the rotated beam of polarised light in figure 4.1 is reflected back down the glass rod, it is rotated further before emerging beside the source. This shows that it is the light itself that is influenced, not the material of the glass.

lead silico-borate glass along the direction of a magnetic field, then the direction of polarisation was rotated (figure 4.1). This effect is now universally known as Faraday magneto-optic rotation or simply the Faraday effect (although he discovered many other effects too).

He soon showed that the same effect was present, although to a lesser degree, in other materials, both solids and liquids. Its properties were that the rotation depended on the material used (a characteristic now known as the Verdet constant of the material), on the length of the material within the field and on the strength of the magnetic field itself. But the big surprise was that if the rotated beam was reflected back towards the source, the rotation was not reversed on the return trip and thus cancelled out, but occurred again so that the rotation was doubled (figure 4.2). This may seem paradoxical because light passing the opposite way in the same field is, of course, then being rotated the opposite way. But clockwise as seen from one direction is anticlockwise when seen from the other direction. So a ray whose polarisation has been rotated clockwise (say from 12 to 1 o'clock as seen from behind), and then reflected back by a mirror is now passing the other way and sets off vibrating at 11 o'clock as seen from behind; during the second passage through the glass it will be rotated anticlockwise, thus increasing the earlier rotation to end at 10 o'clock.

Michael Faraday had originally assumed that the magnetic field would somehow stress the atoms within the glass, in which case one would expect any effect on light to depend on its direction of travel and to be 'unwound' again on the way back. But the actual situation suggested that light itself must have some directional property across its line of travel. Faraday said it is as if the glass itself is somehow rotating in the magnetic field and carrying the light round with it. One must remember that the very nature of light was not at all well understood at the time but this discovery was clearly important in giving a new angle on it. Twenty years later, in 1865, James Clerk Maxwell of King's College, London published the now famous Maxwell equations, for which he acknowledged the influence of Faraday and specifically the magnetic rotation of polarised light. The new theory described light as an electromagnetic wave and predicted a much larger class of such waves of other wavelengths. We now know of electromagnetic waves extending from radio, with some wavelengths of over a thousand kilometres, through microwaves, infrared, the visible light spectrum at less than 1 micrometre (μm), ultraviolet, x-rays and, finally, gamma-rays less than 10^{-14} m: a continuous range of over 20 orders of magnitude. All can be described as the same interaction of a magnetic and an electric field, and all propagate at 'the speed of light'.

Faraday rotation is easy to demonstrate although the very dense glass that gives a large effect is extremely expensive. Figure 4.3 shows a homemade device using a 'borrowed' rod of density 6.63 g ml^{-1} (most glasses have a density of around 2.5 g ml^{-1}). It is 1 cm in diameter and 5 cm long and is held in a plastic bobbin wound with thick wire, giving a resistance of half an ohm. A polaroid is fixed across one end and a rotatable polaroid cap on the other end is crossed with the first to give extinction of the light. Passing current from a 12 V lead–acid battery through a heavy-duty push button then allows light to shine through brightly. For demonstration to an audience, the device is put over a hole in a cardboard mask on the patten of an overhead projector. A very similar device, though not using polaroids of course, that was made by Michael Faraday himself is on display in the Faraday Museum at the Royal Institution in London.

In one of his papers, Faraday said that fused lead borate gave an

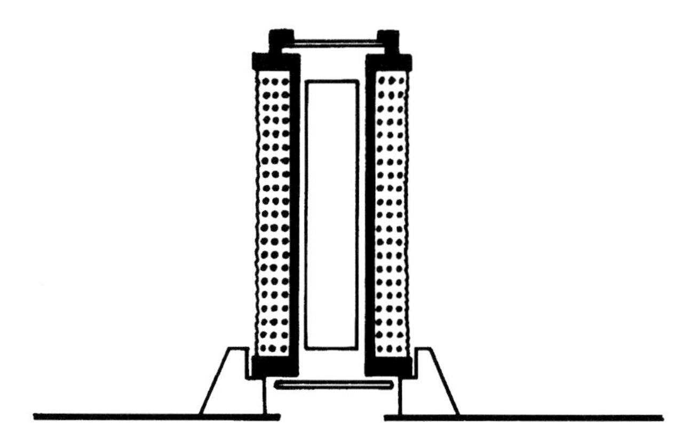

Figure 4.3. A Faraday rotation device for demonstration on an overhead projector. As it draws about 25 A from a 12 V accumulator, the power can only be maintained briefly, so a push button is used instead of a switch. When 'fired' it allows a bright beam to pass through crossed polaroids, thus demonstrating an extremely important historical phenomenon. The effect is made more clearly visible if the device stands over a hole in a cardboard mask.

effect that was equal to the best heavy glasses he tried. Two of my colleagues kindly helped me to try this. John Cowley, a glassblower, made a rod of fused lead borate with materials and a mould supplied by Isaac Abrahams. This rod is 55 mm long and has a density of about 6.25 g ml^{-1}. It is now mounted in its own coil (as in figure 4.3), using 500 g of enamelled copper wire 1.35 mm in diameter. It gives excellent results for direct viewing or for demonstration on an overhead projector, although the rod is coloured greenish rather than being clear.

The principle of the Faraday effect can be used to make a true one-way system for light—an optical isolator, invented by Rayleigh in 1885. The principle of this is shown in figure 4.4. Linearly polarised light is rotated by 45° by a magnetic field so that it passes through a second polariser orientated obliquely at the appropriate angle. But light passing in the opposite direction and starting with polarisation at 45° is then rotated further so that its exit polariser then appears crossed and extinction occurs.

Faraday rotation is not just a laboratory effect for it is used in astronomy to measure the strength of magnetic fields in the space between stars. Where it is possible to infer the original polarisation of

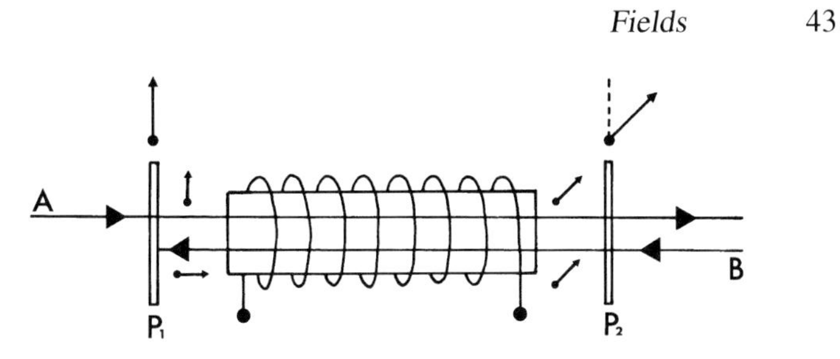

Figure 4.4. The principle of a one-way optical isolator. The left polariser P_1 is here set vertical; the light passing through it from A is then rotated clockwise by 45° by a Faraday rotator cell and passes freely through the right polariser P_2, set obliquely at 45°. But the lower beam of light entering from B is first polarised at 45° and then rotated to become horizontal so that it is blocked by the left polariser. The device is therefore transparent in one direction and opaque in the opposite direction. Large vector arrows show the setting of the polarisers, small ones show the polarisation of each light beam.

a light source (say when it is produced by scattering—see chapter 6), then any change in the direction of vibration during its journey to earth indicates the magnetic fields in the intervening space. In this way it has been shown that the interstellar magnetic field within our galaxy is about one-millionth of the strength of our own terrestrial magnetic field and is aligned along the spiral galactic arms.

Michael Faraday also tried to influence light with electric fields but he never succeeded. In the previous quotation, his use of the word 'electrifying' clearly refers to electric currents through the electromagnets used in some of his experiments, and not to a direct effect of electricity on light. It was not until 1875, 10 years after Maxwell's theory and 3 years after Faraday died, that this was achieved by John Kerr. A Kerr cell (figure 4.5) consists of two sets of small plate electrodes immersed in nitrobenzene. They are set diagonally between crossed polaroids. An electrical pulse of several thousand volts then twists the direction of vibration so that the light can pass the second polaroid. Such a device, which is then sometimes called a Karolus cell, is used as the shutter for very high speed cameras because its action is extremely rapid. With a suitably short pulse of high voltage, the cell can be opened for as little as 1 ns (10^{-9} s), or one-thousandth of a millionth of a second. The light pulse that passes will then be only about 30 cm long—less than 1 foot! It may help to visualise such a short duration by saying that,

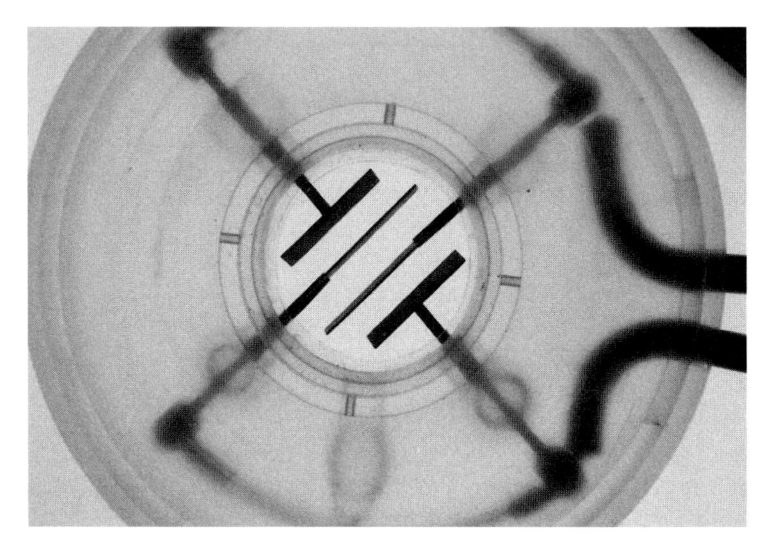

Figure 4.5. A Kerr cell showing the electrodes that are connected together alternately in pairs and immersed in nitrobenzene. The electrodes are set at 45° to the vibration directions of two crossed polaroids. A momentary application of several kilovolts rotates the direction of polarisation so that the combination becomes transparent. It can be simply demonstrated on an overhead projector, using a piezo-electric gas-lighter as the high voltage source.

going the other way, 10^9 s is nearly 32 years. Such very short exposure times allow sharp pictures to be taken of extremely fast events such as the course of development of explosions.

In 1877 John Kerr also showed that when polarised light is reflected from the surface of iron or certain other metals, alloys and even some oxides, the direction of polarisation is rotated by the application of a magnetising field. This effect is therefore called the magnetic Kerr effect.

Both Faraday cells and Kerr cells can be used to modulate the intensity of light for communication through light fibres. The currently fashionable form is a Pockel cell which is like a solid state version of the Kerr cell that works in a crystal, often potassium dihydrogen phosphate, instead of in liquid nitrobenzene. The physical basis of this device is somewhat different in detail but the effect is virtually the same. It can vary the light entering the optical fibre so rapidly that up to 100 000 telephone calls or 100 television channels can be transmitted simultaneously down a single tiny fibre.

So Michael Faraday's experiments in this area alone led to profound insights into the very nature of light, to a method for exploring features of the galaxy, to the fastest camera shutters and to the latest techniques for high-speed communications. There really is no way of knowing where a discovery may lead, no matter how 'rarified' it or the work leading to it may seem at the time.

Chapter 5

Left hand, right hand

Some chemical solutions rotate the direction of vibration of polarised light despite the fact that their constituent molecules are randomly orientated and are continually jostled by thermal agitation. There is no trace of a regular lattice in this case and there is no need for a permeating field. All that is necessary is that the molecules of the dissolved substance (the solute) can exist in both 'left-handed' and 'right-handed' forms and that one of them predominates in the given solution. Just as gloves, and hands, come in mirror image pairs (a left and a right), some molecules come in left- and right-handed forms and their property of handness is called chirality. The word chiral comes from the Greek *cheir* ≡ hand.

The explanation of the effect of molecular handedness on polarised light began with another brilliant piece of mid-19th century research, this time by Louis Pasteur who was originally a chemist, although perhaps better known later as a pioneer of microbiology and vaccination. In 1848, when he was only 26 years old, he studied the chemistry of tartaric acid (a salt of which is cream of tartar) which had been produced from wine residues (or tartar) and was known to rotate polarised light in solution. Another acid, called racemic acid, that had been discovered in 1820, was found to have the same formula, $C_4H_6O_6$, as tartaric acid and its other properties were also identical except that it did not rotate polarised light at all. Tartaric acid was said to be 'optically active' and 'dextrorotatory' (twisting polarised light to the right—or clockwise as seen from in front) whereas racemic acid was optically inactive. Indeed when tartaric acid is heated in water it slowly loses its optical activity without any apparent change in molecular structure

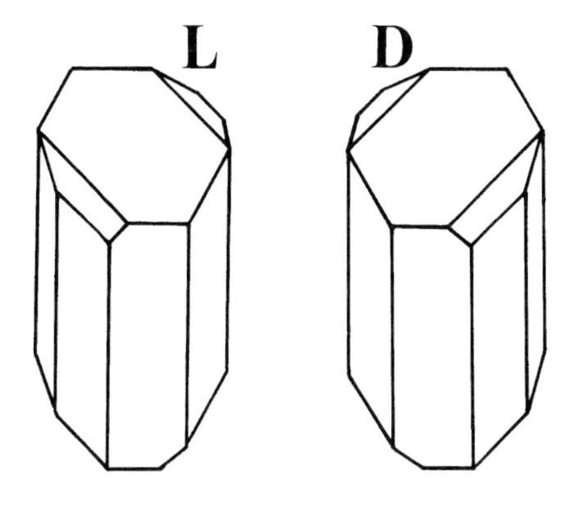

Figure 5.1. A crystal of sodium ammonium tartrate (left) which is laevorotatory (L) and its mirror image (right) which is dextrorotatory (D). Pasteur showed that equal numbers of each kind are found in sodium ammonium racemate which is therefore optically inactive. (There has been much confusion over the proper shapes of these crystals and different shapes are often depicted; these patterns are copied from Pasteur's own paper although one authority has declared that Pasteur never published any figures of the crystals he described!)

and becomes racemic acid. The two substances were even thought (in this case wrongly as it turned out) to produce identical crystals which made it even more puzzling.

But Pasteur found that there were tiny differences in the crystals. When he took racemic acid and prepared crystals of the salt sodium ammonium racemate, he noticed that half of them were actually mirror images of the rest (figure 5.1), and one of these matched the crystals of sodium ammonium tartrate. Although they were all very small, Pasteur carefully sorted the racemate crystals into two piles, using a magnifying glass and tweezers. When he dissolved them separately in water, both were optically active; one solution rotated polarised light to the right like the corresponding tartrate, while the other rotated it to the left. The latter was something not known before: a left-handed or 'laevorotatory' tartrate. The amount of rotation was equal in each case and when the solutions were combined they counteracted each other exactly to produce the typical optically inactive racemate. Pasteur then surmised that optical activity must be related to the fact that the crystals were mirror images. Furthermore, since optical activity persisted when the crystals were dissolved in water, then even the molecules of tartaric acid and its salts must themselves exist in two forms which are mirror images of each other. We now call such pairs of molecular structures stereoisomers and an optically inactive mixture of the two is said to be racemic.

So surprising was this result that Pasteur, then so young and inexperienced, was invited to repeat his procedure under the supervision of an eminent old scientist called Jean Baptiste Biot who had himself discovered optical activity in liquids 30 years before. All went well and Pasteur's reputation was established with a discovery that was to have immensely important consequences for the understanding of molecular structures. But it might easily have turned out very differently. Sodium ammonium racemate only forms two different kinds of crystal at temperatures below 26–28 °C; above that (like most other chiral salts, including other tartrates) the two stereoisomers form combined, or racemic, crystals that are quite different in their appearance and properties. It seems that Pasteur did not know this at the time, and if there had been a heatwave when he repeated the experiment it might not have worked, which would have been disastrous to his reputation. Indeed in hot weather he might not have made his original discovery at all.

Although Pasteur deduced that mirror image crystals must represent a handedness in the shape of their constituent molecules, the details eluded him. Twenty-six years later, in 1874, two chemists called van't Hoff and Le Bel independently explained how some molecules that contain carbon atoms can exist in two forms that are optically active mirror images of each other. For a molecule to be chiral it must contain at least one atom that is joined to four different atoms or groups (figure 5.2) to form a three-dimensional structure (it does not work if they happen to be all in the same plane); then there are two possible configurations, or stereoisomers, that are mirror images and cannot be superimposed, one on the other. Now a carbon atom makes four links to other atoms as if they were attached to the corners of a tetrahedron, so chiral molecules are quite common in organic chemistry and therefore in living materials and natural products.

When only one of the stereoisomers is present (or when one predominates) in a solution, then it acts to twist the vibration direction of polarised light passing through it. The amount of rotation depends on the characteristic 'specific rotation' of the substance, on the concentration of the solution, on the temperature and on the length of light path through the solution. Shorter wavelengths of blue light are generally rotated more than the longer wavelengths of red light. If the rotated beam is

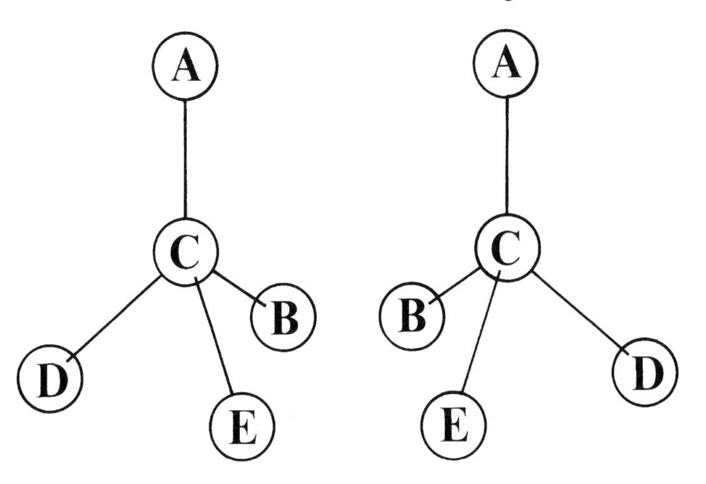

Figure 5.2. Two hypothetical three-dimensional molecules with a carbon atom C joined to four different atoms or groups of atoms, A, B, D and E. Although they have the same formula, it is impossible to superimpose them and each resembles the mirror image of the other. The components A, B, D and E may be anything from a single hydrogen atom to a long, complex chain provided that they are all different. Each of these 'chiral' molecules will rotate polarised light in opposite directions, even in free solution, although all their other properties are identical.

reflected back through the solution, it is again rotated in the same sense but, as it is now travelling the opposite way, the original rotation becomes cancelled out and the light is unaffected after the double passage. This is in marked distinction to the cumulative rotatory effect of magnetic fields in the Faraday effect (chapter 4).

When a molecule has more than one atom that is linked to four different groups, then there are generally more than two possible configurations: in principle, two such atoms give four possibilities and every extra carbon atom doubles the number of possibilities. Figure 5.3 shows that tartaric acid has two asymmetrical carbon atoms (shown in heavy type and with a central dot) but, as their connections are identical, two of the possibilities turn out to be the same. So there are actually three tartaric acid molecules: one is laevorotatory at both ends (*l*–*l*), one is dextrorotatory at both ends (*d*–*d*) and an equal quantity of these produces an optically inactive racemic mixture (*l*–*l* + *d*–*d*). The third form, also discovered by Pasteur and called mesotartaric acid, is (*l*) at

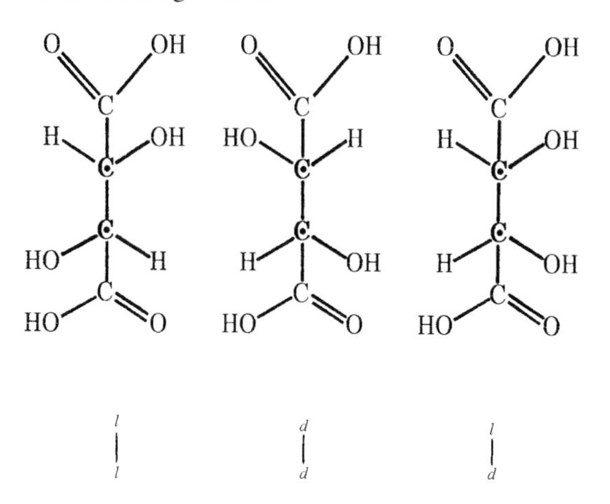

Figure 5.3. Three possible configurations of tartaric acid molecules projected into two dimensions. The laevorotatory (*l*–*l*) form is the mirror image of the dextrorotatory (*d*–*d*) form. The third (*d*–*l*) form, called mesotartaric acid, is (*d*) at one end and (*l*) at the other: it can be superimposed on its own mirror image if it is also turned lengthwise and it is therefore optically inactive and can be regarded as 'ambidextrous'.

one end and (*d*) at the other; there is only one such form because if each end is changed the molecule can be turned end for end to be the same as before. This (*l*–*d*) molecule is optically inactive because the two ends effectively counteract each other.

Another way of looking at it is to say this 'ambidextrous' molecule is not asymmetrically chiral: it can be superimposed upon its own mirror image and therefore cannot be optically active. When tartaric acid is synthesised in the laboratory, all three forms are normally produced because the ends of the molecules are usually generated randomly. This mixture is optically inactive because the racemic mixture of (*l*–*l*) and (*d*–*d*) balances out and the meso (*l*–*d*) form is itself inactive. Under heat treatment, as mentioned earlier, some of the (*d*–*d*) form that is found naturally in grape juice becomes converted to the other two forms until all optical activity is eventually lost. Some other chiral substances will do this spontaneously in a process called autoracemisation.

Since the tartaric acid molecule can exist in three forms, then so can many of its salts—in which hydrogen ions on either end are replaced by other ions such as sodium or ammonium. If both ends are replaced by the same kind of ion, as in disodium tartrate, then three forms are possible: left, right and ambidextrous. But when the two ends have different replacements, then four different forms are possible, since (*l*–*d*) is not equivalent to (*d*–*l*) when the first and second groups are different, and there are also the (*d*–*d*) and (*l*–*l*) forms. In principle, this is the case with sodium ammonium tartrate which Pasteur studied, but this complication did not arise in his case. The two 'quasi-ambidextrous' forms can only be prepared from ambidextrous mesotartaric acid and this does not occur in natural grape juice. So Pasteur was only dealing with the left-handed (*l*–*l*) and right-handed (*d*–*d*) forms.

Optical activity is also a property of many sugars and these form a splendid basis for demonstrating the phenomenon with the simple apparatus shown in figure 5.4. A glass cylinder with a flat bottom stands over a polaroid and is capped by a second polaroid that is free to rotate. A radial pointer is fixed to the upper cap to show its orientation and the whole apparatus can be mounted over a hole in a cardboard mask so that an overhead projector can show it to an audience (it helps if the projector lens can be raised so that it focuses at about the middle of the tube).

With water in the tube, the two polaroids are crossed so that no light passes, and the position of the pointer is noted (perhaps on a simple dial). Then the water is replaced by a strong solution of cane sugar (sucrose); light now comes through and the upper polaroid must be rotated to the right (clockwise) in order to get extinction once again. The sucrose is seen to be dextrorotatory. This apparatus is a simple form of a polarimeter, also called a saccharimeter in the sugar refining industry where it is routinely used to measure the strength of sugar solutions. The new extinction is not sharp but changes colour over a small range without going black, due to the dependence of rotation on wavelength, where blue light is generally rotated more than red light. It can be made sharper and more convincing if a piece of coloured cellophane is added as a filter. Although blue shows the greatest rotation, green is almost as (and sometimes rather more) effective and is usually brighter.

Glucose (grape sugar) is also dextrorotatory, but fructose (fruit sugar) is laevorotatory so the upper polaroid must be turned to the left (anticlockwise) in order to restore extinction. For this reason glucose (a *d* sugar) is often called dextrose and fructose (an *l* sugar) is also called laevulose. Like characters in a Russian novel, many sugars have three names! Now the sucrose molecule is exactly like a glucose molecule and a fructose molecule joined together. In the presence of a specific enzyme called sucrase (or invertase), or when boiled with a little acid,

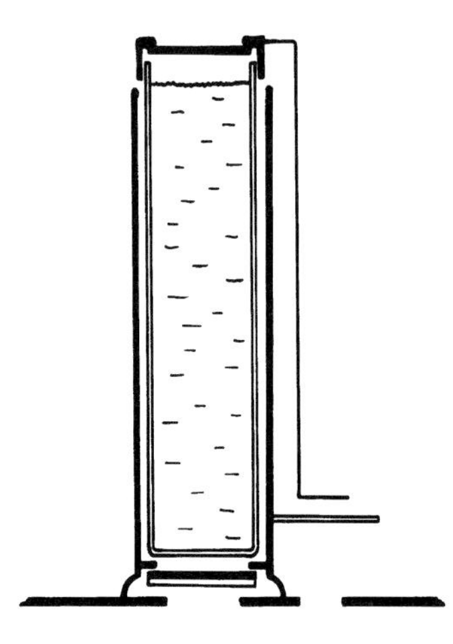

Figure 5.4. A simple polarimeter for demonstrating the optical activity of sugar solutions. A glass column of solution is capped at each end by polaroids (shown black) that can be rotated to extinguish the passage of light from below. When sugar solutions are present, the setting of the caps must be changed. Sucrose and glucose are *d* sugars, twisting the direction of polarisation clockwise; fructose is an *l* sugar that twists the vibration anticlockwise. A cardboard mask helps when the apparatus is shown on an overhead projector and if possible the projector should be focused about halfway up the tube. A radial wire pointer may then be brought down to read against a transparent scale, illuminated by a second window in the mask as shown on the right. Extinction is sharper if a green filter is placed under the column.

the sucrose combines with a molecule of water and splits into its two components:

$$\underset{\text{(sucrose)}}{C_{12}H_{22}O_{11}} + H_2O = \underset{\text{(glucose)}}{C_6H_{12}O_6} + \underset{\text{(fructose)}}{C_6H_{12}O_6}\,.$$

Although glucose and fructose have the same basic formula, their internal structures are different (figure 5.5) and they do not form an inactive racemic mixture. Fructose is inherently much more optically active than glucose (as shown by the numbers in table 5.1), so the equal mixture turns polarised light to the left. The right-turning sucrose is said to have been inverted or changed into 'invert sugar'. This is often done in wine and beer making as it is then more readily fermented. Invert sugar can be bought ready to use but yeasts produce their own sucrase so

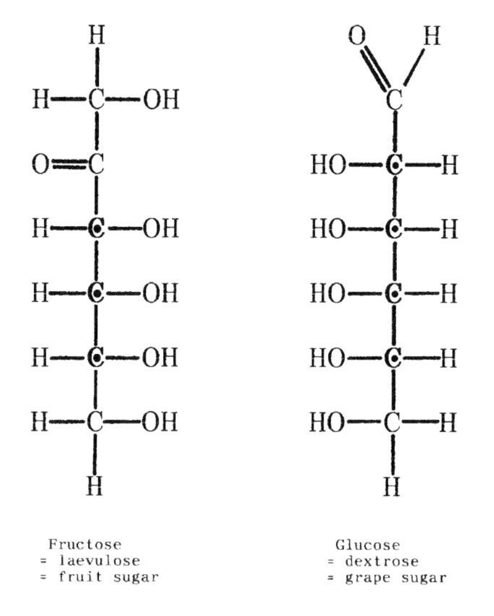

Figure 5.5. The structural formulae for fructose and glucose projected into two dimensions. Heavy type and a central dot show the asymmetrical carbon atoms, each connected to three different atoms or groups, of which fructose has three while glucose has four. In principle, there can be eight forms of fructose and 16 forms of glucose but only one of each occurs naturally.

they can manage quite well, only a little more slowly, to ferment sucrose after inverting it themselves. For a simple demonstration, domestic cane sugar can be boiled for some minutes with a little citric acid (or vinegar etc) and examined in the polarimeter before and after. Glucose is readily available and fructose can be bought at health food shops since it is very sweet but less harmful to diabetics than the usual culinary sucrose.

Figure 5.5 shows that fructose has three asymmetrical carbon atoms and therefore can exist in eight different configurations, while glucose has four asymmetrical carbons giving 16 possibilities. Half the forms of each sugar will be dextrotatory and half laevorotatory but only one form of each is ever found in nature. Actually, when glucose is first dissolved in water it shows a high optical activity that then gradually declines to less than half (from +112 to +52.7) over a period of time. Fructose also changes but to a lesser extent (from −132 to −92.7). In both cases this is because the molecules change gradually by spontaneous rearrangement of internal linkages from the form they have in the crystals to a somewhat

Table 5.1. The specific optical activity (as measured under specified conditions) for sugars mentioned in the text. A positive value indicates rotation to the right, a negative one to the left.

Sucrose (cane/beet sugar)	+66.5	
Invert sugar	−19.4	
d-glucose (grape sugar, dextrose)	+112	+52.7
l-fructose (fruit sugar, laevulose)	−132	−92.4

less active form in free solution. This behaviour explains why two values are often given for their optical activities in reference tables, including table 5.1. The molecular structures and therefore the optical activities can also be markedly influenced by the physical conditions of the solutions, such as temperature.

All other natural sugars also occur in only one of their possible chiral forms. Some are (*l*) sugars and some are (*d*) sugars but they can all be derived from a very simple sugar called *d*-glyceraldehyde which has the formula $C_3H_6O_3$ and just one asymmetrical carbon atom. No natural sugars that can be derived from *l*-glyceraldehyde are known. For this reason natural sugars are said to be right-handed D-sugars (with a capital D), from the dextrorotatory series, regardless of their own (*l*) or (*d*) optical activity. The corresponding L-sugars do not occur naturally.

Amino acids are also very important components of all living systems since they are the building blocks of all proteins including enzymes. Of the 20 amino acids that can combine to form proteins, 19 are optically active (one is non-chiral) and all are related to *l*-glyceraldehyde; they are therefore said to be all left-handed or L-amino acids (with a capital L). The corresponding D-amino acids are virtually absent from nature. Furthermore, the complex proteins formed by combinations of amino acids have a variety of characteristic shapes. The common alpha-helix, when formed from L-amino acids is always right-handed (like a right handed screw—figure 5.6) so that many structural elements in our bodies are of only one chiral form. Even DNA, which carries the genetic code, is coiled into a right-handed double helix because it is composed of D-sugars and L-nucleotides (figure 5.6).

[The handedness of a helix causes much confusion due to

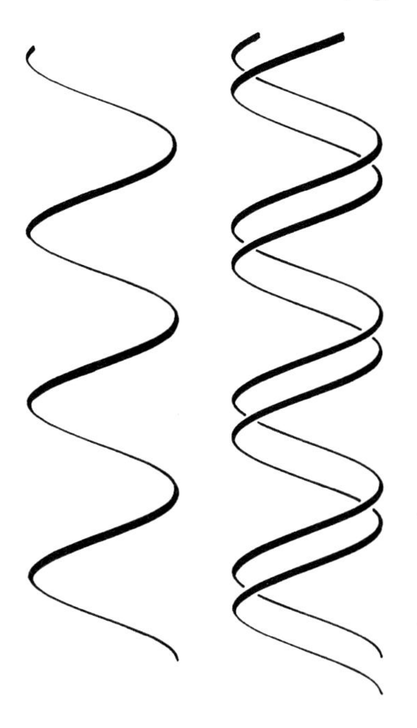

Figure 5.6. Left: the shape of the right-handed alpha helix, which is a common configuration in proteins and forms naturally from L-amino acids; the mirror-image left-handed helix would need D-amino acids which do not occur naturally. Right: the double helix of DNA is also right-handed due to the chirality of its constituents.

its ambiguity: if it is clockwise or 'right-handed' when viewed from behind and traced away from the observer, as with most fixing screws, then it is anticlockwise or 'left-handed' when viewed from in front and traced towards the observer. Different conventions are used in different fields or even in different countries: the rear view convention seems to be commoner although botanists use the front-view convention to describe twining plants—a 16th century book on the cultivation of hops (regarded as twining to the right) said they should be trained round their poles 'alwayes according to the course of the sunne', although this advice would obviously be wrong in the southern hemisphere where the sun's course appears reversed. Even illustrations, which might seem to be unambiguous, may actually add confusion: the alpha and double helices are often depicted the wrong way round, suggesting that the artist was confused by the instructions given. This book uses the rear-view convention for helices except for optical rotation where there appears to be universal acceptance of the front-view convention (as seen by the user of a polarimeter) so the latter is used here for this topic only.]

So all life 'as we know it' is based upon D-sugars and L-amino acids and the reason for this is an outstanding puzzle. Several theories have been proposed to explain how this came about but so far none has proved entirely convincing. A recent idea is that it might have been connected to the processes of radioactive decay which are inherently asymmetrical and produce particles that spin in only one direction. It has recently been found that left-spinning electrons can bias the formation of chiral crystals of sodium chlorate (see later), which are otherwise produced from solution in equal numbers. But maybe the handedness of biological molecules was originally due to some completely random event when life was only beginning on earth.

It is much easier to understand why only one form should take precedence over the coexistence of both. Food is broken down by enzymes, which are proteins whose overall shapes are important for their function because they must 'fit' onto the molecules they work on. Their overall shapes are determined by the shapes of their constituent amino acids. Enzymes composed of D-amino acids would themselves be mirror images of the ones made of L-amino acids that we find today, and they would probably not work well if at all because they would not fit their substrates in food, which are themselves chiral. So life based on L-sugars and D-amino acids is perfectly conceivable but would be incompatible with the life forms here on earth.

At the beginning of *Through the Looking-Glass*, Alice speculates on the presence of milk for her kitten in Looking-glass House, but she decides that looking-glass milk might not be good to drink. In 1872 Lewis Carroll could hardly have known that, because of chirality, the mirror-image constituents of looking-glass milk probably would be indigestible and of little use to real-world kittens. One can also imagine deep-space explorers of the future arriving at a beautiful planet, just like earth in most respects, but being unable to replenish their stores because on that planet life is based (by chance?) on L-sugars and D-amino acids and food substances there would be of little or no nutritional value to earthmen. (Strangely perhaps, stereoisomers may smell quite different from each other.) If one imagines that an enzyme fits a part of the protein it attacks like a glove fitting on a hand, then feeding proteins made of D-amino acids to any creature from Earth would be like giving a supply of right-handed gloves to a person with only a left hand: they might be beautifully made and all just the right size, but completely useless! Of course if the radioactive decay theory for the origin of chirality is correct, then all life everywhere in the Universe would have the same handedness and so be compatible after all. This is another example of an apparently simply observation (the rotation of polarisation by some solutions) leading to conclusions that are very profound indeed.

At present there is much interest in the development of chiral pharmaceuticals, or drugs whose molecules are all of the same handedness. Many natural chemical agents within the body are chiral and match the receptor sites at which they act. So any drugs intended to act at these sites must also be chiral. If a racemic mixture is administered, then at best the 'unwanted' stereoisomer is inactive and simply represents a waste of manufacturing potential. But in many cases the 'wrong-handed' molecules cause side effects in quite different areas. The most infamous example is that of thalidomide which was used to alleviate morning sickness during early pregnancy. This drug is chiral and only one stereoisomer is effective; unfortunately the other stereoisomer was found, too late, to cause very severe defects in embryonic development, leading to babies with greatly reduced or missing limbs. Now other uses have been found for the 'good' isomer and there is pressure to produce it in a pure form so that it can be administered safely—although spontaneous racemisation may be a danger in some cases.

When chiral chemicals are synthesised from non-chiral components, both isomers are produced giving a racemic mixture. Selective production of one stereoisomer can be achieved in several ways, all essentially originated by Pasteur, that are now being exploited industrially, for example by ChiroTech of Cambridge. Some molecules can be synthesised in only one chiral form by using enzymes, either *in vitro* or by the activity of living micro-organisms such as bacteria or yeasts; since enzymes are chiral, any of their chiral products will be all of one stereoisomer. Other asymmetrical catalysts are now available to allow asymmetrical syntheses. Single stereoisomers can also be extracted from racemic mixtures by selective crystallisation. A saturated solution is seeded with tiny crystals of one stereoisomer and the resulting crystals, all of one kind, are collected; then seeds of the other stereoisomer are used and the resultant crystals discarded. By alternating in this way a high yield can be obtained with very low contamination. Seeds can be very small so a single crystal, once isolated, can be fragmented and multiplied quickly.

Sometimes a racemic mixture can be combined with one stereoisomer of another chiral compound; the two kinds of molecule thus formed will not be simply mirror images but will have quite different shapes. This means they will have different solubilities which makes them easier to separate by crystallisation before the wanted stereoisomer of the original substance is recovered by separation from its temporary partner. Pasteur managed to separate the stereoisomers of racemic acid in this way by first combining them with naturally chiral quinine.

Another naturally occurring, optically active substance is turpentine. This is a mixture of chemical substances obtained by 'bleeding' the trunks of various coniferous trees. The direction in which it rotates polarised light varies, depending on the species of tree from which it is derived; it is commonly laevorotatory but may be dextrorotatory. Either way, this forms a simple test for 'genuine' turpentine since the cheaper synthetic turpentine substitute is optically inactive.

One further occurrence of chirality must be mentioned—that found in certain crystals. One example is sodium chlorate ($NaClO_3$) where the molecules are not chiral so that a solution in water is optically inactive. But when crystals form, although they have simple shapes (generally cubic but under some conditions tetrahedral), their internal lattice structure may take either of two mirror-image patterns, so that any given crystal will be either laevorotatory or dextrorotatory.

Another example is quartz. In chapter 3 it was stated that crystals show no birefringence for light passing along their optical axes. This remains true for quartz but quartz crystals do show optical activity, rotating the direction of polarisation, for rays along their axes. Quartz is silicon dioxide and the SiO_2 molecule itself is not chiral so that fused quartz is optically inactive (it is of course insoluble in water). But the lattice of quartz crystals consists of a complex mesh of interlocking helices (often, wrongly, called spirals), with each atom being incorporated in several helices with different axes. A helix is chiral and, once a crystal starts to grow, it must be consistent, so that quartz crystals come in two chiral forms which are mirror images of each other (figure 5.7) and show the opposite optical activity along their axes. So in

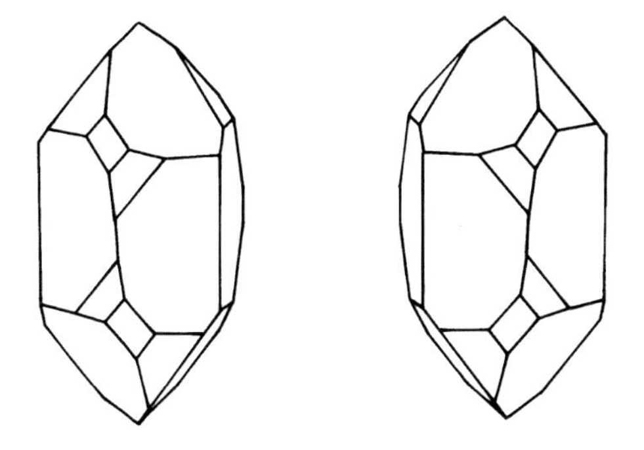

Figure 5.7. Quartz occurs in two chiral forms and in alpha quartz these can be distinguished on a macro scale by small facets; these facets are missing from beta quartz which is more difficult to characterise chirally. Polarised light passing along their optical axes is rotated in opposite directions in each kind so that they can be designated *laevo* and *dextro* quartz, here seen on the left and right respectively.

both quartz and sodium chlorate, the molecules themselves are not chiral and chirality is a property only of the crystal lattices.

Slices of quartz crystals cut normal to their optical axes are commonly used as elements in optical instruments (including some designs of sensitive polarimeters), often as left- and right-handed pairs side by side. Because different wavelengths are rotated to different degrees, a thick slice of quartz cut normal to its optical axis shows colours between crossed polarisers that change when either polariser is rotated—but not when the crystal itself is rotated. These 'rotation colours' are therefore quite distinct from the colours produced by the property of birefringence that retards light crossing the optical axis of the same crystals as described in chapter 3. Quartz crystals are much easier to identify, and may be much larger, than the tartrate crystals separated by Pasteur, but the matter is complicated because quartz occurs in two forms depending on the temperature at which it was originally grown. 'Alpha' quartz has a distorted lattice and small extra facets on the 'shoulders' of its column that make its chirality obvious (figure 5.7) but these are missing from the common 'beta' quartz. Also real quartz crystals are often distorted and seldom look quite like the pictures 'in the books'. Only perfect alpha quartz is easily identified in this way.

Chapter 6

Scattering

On the moon, the sky is black and the stars shine clearly even when the sun is high. But here on earth the clear daytime sky is blue and no stars can be seen through it. This was explained by John Tyndall, Director of the Royal Institution, in the 1860s as the scattering of sunlight in the upper atmosphere. He showed that small particles scatter light of short wavelengths more strongly than longer ones: blue more than red. That is why fine smoke and mist look blue. It is well known that distance, as depicted in a painting for example, involves a graded blue tinge. A. E. Housman asked 'What are those blue remembered hills?' and Thomas Campbell gave the answer ''Tis distance lends enchantment to the view, And robes the mountain in its azure hue'. Tyndall explained that even on clear days a very long propagation path will accumulate shorter wavelengths that have been scattered to form a blueish veiling.

Lord Rayleigh soon worked out the theory of it and calculated that even gas molecules in sufficient quantity will scatter blue light, so that air in the upper atmosphere looks blue without necessarily containing solid or liquid particles. For this reason molecular scattering is often called Rayleigh scattering. He found that for very small particles, smaller than the wavelength, the amount of light scattered is inversely proportional to the fourth power of wavelength, so that deep blue light of 420 nm is scattered ten times more strongly than deep red light of 750 nm. It follows of course that shorter wavelength ultraviolet (UV) light is scattered even more strongly, so the sky is very bright in the near-UV, even though UV is less strong than yellow light in sunlight itself. Many animals can see this near-UV although we humans cannot.

Tyndall scattering by small particles is very common in nature, and some familiar examples are the colour of blue eyes, the blue faces and rear ends of mandrills and some monkeys and the blue feathers of many birds such as tits and kingfishers. Wild-type green budgerigars combine Tyndall scattering with an overlying yellow filtering pigment to give their bright green colour. A genetic deficiency in producing the yellow screening gives the blue varieties, and a failure to develop the Tyndall scattering particles gives yellow birds. A similar combination is found in green tree frogs.

John Tyndall performed laboratory experiments on the scattering of light by smokes in air within a long glass tube, part of which is still on display at the Royal Institution. He found that the scattered light is polarised in a direction normal to the axis of the beam. In a horizontal beam of light all vertical components are scattered sideways (horizontally) and horizontal components are scattered upwards and downwards (vertically: figure 6.1). The degree of polarisation is greatest, nearly 100%, when viewed 'side-on' at 90° to the original beam and it becomes less as the angle of scattering decreases both forwards and backwards. All this was given a theoretical basis by Lord Rayleigh and it explained the fact that blue sky light is polarised, as had been observed by the Frenchman François Arago back in 1809. The actual degree of polarisation is always less than the 'theoretical' value, often markedly so, due to secondary rescattering of the light on its way to the observer. Blue eyes also show changes in brightness when viewed through a rotated polaroid. Irises of other colours have larger scattering particles and can show some polarising effect provided they are not too dark.

A beautiful demonstration of polarisation by scattering can be made with a column of water held above a bright halogen lamp (figure 6.2). A few drops of Dettol (or milk) disperse as very fine droplets to form a pale blue scatterer (colour plate 22), and a piece of polaroid shows that this light is polarised. If the polaroid is placed under the column, then the light is already polarised and can only be scattered strongly in two opposite directions and not at all at right angles. This is seen by the naked eye without the need for another polariser since the column itself acts as the second polariser. Also the polarisation, and therefore the brightness, is strongest for light that is scattered sideways at right angles to its own path; if the column is viewed at smaller angles to its axis, the polarisation is progressively weaker and dimmer as indicated in figure 6.1. Turning the polaroid then rotates the two scattered 'beams' round the room so that the intensity of scattering appears to wax and wane.

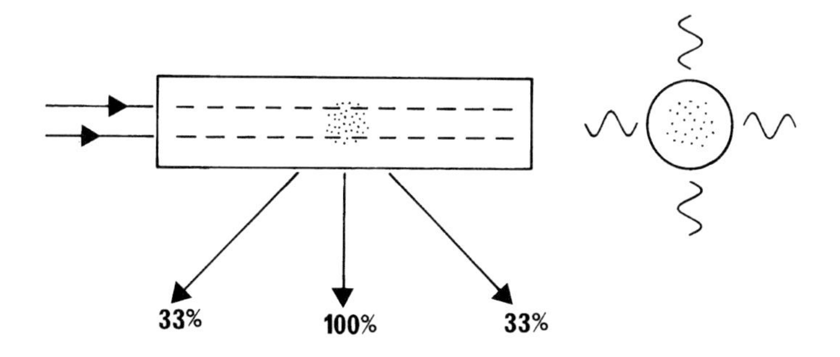

Figure 6.1. Light passing through a suspension of fine particles, such smoke in air, becomes scattered. Light scattered at right angles is essentially 100% polarised with the direction of vibration normal to the path of the original beam (the scattered beam shown here will be polarised in a direction normal to the page). Light scattered in more forward or more backward directions is progressively less strongly polarised as indicated, although the predominant direction of polarisation remains the same (only a small scattering region is shown). At the right is a view along the column, showing the polarisation directions for light scattered at various angles around the axis.

An especially splendid effect is given if a retarder film of cellophane (see chapter 2) is put above the polaroid. The scattered light is then deeply tinged around the column, say blue in two directions, front and back, and the complementary orange at right angles to each side, where there was previously no scattered light. Again these effects are seen by eye alone, without another polariser. A different retarder film may give purple and green and so on (colour plate 23). These other colours also show that it is not only blue light that is scattered, although that is the predominant tinge when retarders are not used. Similarly the blue sky shows the whole spectrum when it is viewed through a spectroscope although the shorter wavelengths are brighter and so give the overall impression of being blue. Such impure blue is said to be unsaturated.

Another spectacularly beautiful demonstration was popular in the late 19th century and has been drawn to my attention by Sir Michael Berry. If the column is filled with a strong sugar solution and a polariser is placed below it, then scattering takes place but its direction is twisted along the tube by the chiral sugar (see chapter 5). Furthermore, different colours are twisted at different rates so that the scattering is coloured through the full spectral sequence along the tube (colour plate 24)—a

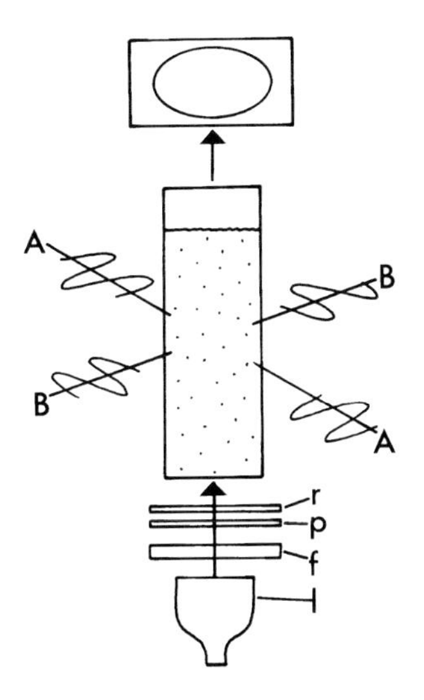

Figure 6.2. A demonstration of scattering by fine droplets (e.g. Dettol or milk) dispersed in a column of water. A strong lamp (l) and heat filter (f) project light upwards into the column. If a sheet of polaroid (p) is placed beneath the column, the light can only be scattered strongly in two directions, say A–A. A retarder film (r) placed above the polaroid colours the scattered light and the complementary colours are then seen to be scattered sideways at right angles to each other: A–A and B–B. Turning the polaroid and the film together makes the colours rotate around the column. Above the column a white card is used to examine the colour of light that is not scattered.

rainbow-hued barber's pole! Viewing at right angles around the axis, or rotating the polariser, shows that at every level each colour is replaced by its complementary colour. If the polariser is rotated slowly and steadily, then the sequence of colours appears to pass smoothly along the column. Quite a short column is adequate if the sugar solution is very strong. Fructose (fruit sugar available from health food shops) is especially effective because it has a much higher specific rotation than sucrose and it is also extremely soluble, so that the syrup can be made very strong.

To return to simple scattering, the properties shown above explain why the polarisation pattern of the blue sky is formed in rings around the sun, weakly near the sun itself but very strongly along the arc at

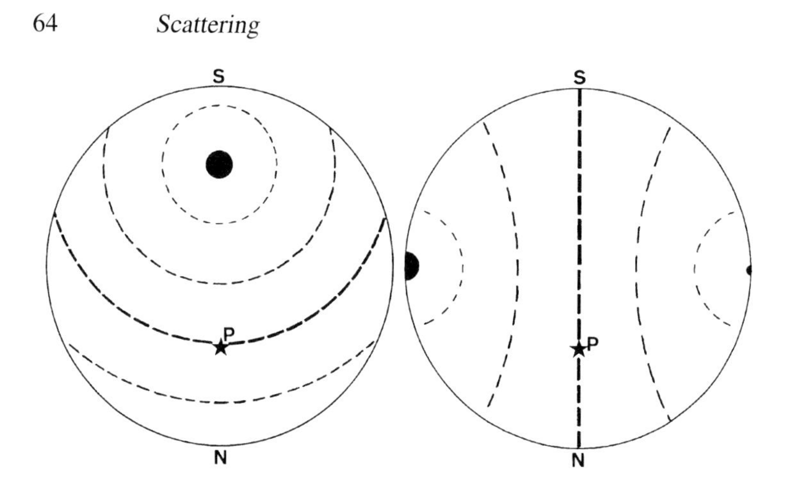

Figure 6.3. Sunlight is scattered by the atmosphere and is polarised in rings around the sun/anti-sun axis. Near the sun the light is scattered forward and is weakly polarised; the arc at right angles to the sun scatters at 90° and is very strongly polarised, while backscattered light from near the anti-sun point is again weakly polarised. The patterns shown here are (left) noon at the latitude of southern England at the equinox and (right) sunset or sunrise. The arcs shown are at 30° intervals and the idealised values for their degree of polarisation are 14%, 60%, 100% and 60% again. Actual values would all be lower due to secondary scattering and there are additional effects, not depicted here, near the horizon. P labels the position of the Pole Star around which the polarisation pattern rotates daily.

right angles to the sun (and its rays) and then progressively weaker again towards the 'anti-sun point' (figure 6.3). The maximum degree of polarisation is rarely more than about 70–80% because some of the light becomes scattered again on its way down to the ground, especially if the air is hazy. Rescattering also adds some slight complexities to the pattern itself, especially near the horizon, but these are seldom significant. The demonstration in the scattering column can also simulate a red sunset. Adding more Dettol scatters more of the predominantly blue light so that what is left, emerging from the top of the column, is orange or red (colour plate 22). Similarly, light from the low setting sun has passed over the heads of people to the west and made their sky blue, so the remaining light is left richer in the longer red wavelengths. Enhanced effects are often produced when the atmosphere contains fine ash from distant volcanoes. It is also true, if rather unromantic, that some of the most spectacular sunsets nowadays occur in cities where scattering is enhanced due to pollution products in the atmosphere.

The polarisation of the blue sky can be very important in nature. Fifty years ago Karl von Frisch was studying the navigation abilities of bees and discovered the way in which they can 'tell' other worker bees where there are rich sources of food. In both cases bees refer to the position of the sun. When a bee flies out to forage, it notes the position of the sun in the sky and uses the reciprocal bearing to find its way back to the hive afterwards. Due allowance is made for the fact that the sun moves across the sky during the day. When food is scarce, a bee may fly as much as 5–10 km from the hive and be away for up to 4 hours, during which the sun moves 60°, yet the insect still makes a 'bee-line' home. It then performs a ritual dance in the hive in which the direction of a good food source is indicated to other bees, again by reference to the direction of the sun.

But von Frisch found that when the sun is hidden by a cloud or behind a mountain or trees, the bees carry on normally. As described in more detail in chapter 9, he found that they are able to detect and use the pattern of polarisation in the sky. We now know that for bees and many other insects the 'sun compass' is actually less important than the 'sky vault compass', formed by to the pattern of sky polarisation which is, however, centred on the sun itself.

In fact the sky polarisation is generally sensed by insects in the near ultraviolet part of the spectrum. Those facets in the dorsal rim region of a bee's compound eye face upwards or forwards (figure 9.5) and so look at the sky. They are also the facets that are sensitive both to ultraviolet and to polarisation. So to a bee the sky must look very different indeed from the sky that we see; instead of being plain blue it must be very bright in the ultraviolet and have a highly structured pattern of polarisation. More details of the sensitivity to polarisation in the eyes of bees and other insects are given in chapter 9.

In recent years it has even been suggested that migrating birds might use sky polarisation as an aid to navigation. Long flights often start around dusk, and for some time after sunset both the band of maximum polarisation and the direction of vibration stretch overhead across the sky in a north–south direction (figure 6.3). As this fades, the stars come out and birds are certainly known to use star patterns as a sky compass.

The sky compass has also been used by man for navigation. The Vikings were intrepid voyagers across the North Sea and northern

Atlantic as much as two centuries before the magnetic compass became available. Between about 860 and 1010 AD they colonised Iceland, discovered and colonised Greenland and even visited both North America and Spitzbergen. Yet the magnetic compass almost certainly did not reach Europe until about 1200 AD. It is widely accepted that the Vikings used the sun and stars in their navigation but cloud would have been a problem. In the sagas there are passing mentions of a 'sunstone' by which the pilot could determine the position of the sun even through clouds. Admittedly the evidence is slim and it has been criticised because it would be of no use without also knowing the time of day, which can itself be measured from the sun only if one knows its direction. But the circularity of this argument can much too readily be used to dismiss the sunstone hypothesis. The passage of time is entirely predictable and can be estimated fairly accurately by the body's 'internal clocks'. Although wind and wave directions are often used to sail a constant course, they may change quite quickly and unpredictably and the use of a sunstone could have drawn attention to this. Any additional estimate of direction, even if not very accurate, might be very useful in an otherwise featureless environment.

Furthermore, the long expeditions to the west and back took many months, so the voyages must have begun as soon as winter was safely over and may not have ended until it was closing in again. At the high latitudes involved, in spring and autumn there would have been virtual twilight for most of the 24 hours with little or no actual sighting of the sun even in fine weather. The sky vault polarisation pattern would have had much to offer at such times, even under clear skies. The sunstone story is also rejected by many authors on the grounds that it would not work at all when the sky was generally overcast. The claim in Rodulf's Saga, in the *Flateyarbok*, that the sun was thus located during a snowstorm is probably a romantic exaggeration—or even bluff as it was done at royal command with no possibility of verification or contradiction—but the technique can be made to work under quite thick cloud cover as the following can show.

It is suggested that the sunstone might have been the pleochroic crystal cordierite or possibly andaluzite, epidote or tourmaline, all of which occur in Scandinavia. It would certainly not work adequately with calcite (Iceland spar) as has sometimes been assumed. A rough, unpolished cordierite crystal mounted in a short tube (figure 6.4) forms an excellent polariscope for observing sky polarisation. The tube shields the crystal from all but one region of the sky and the user should look *at*

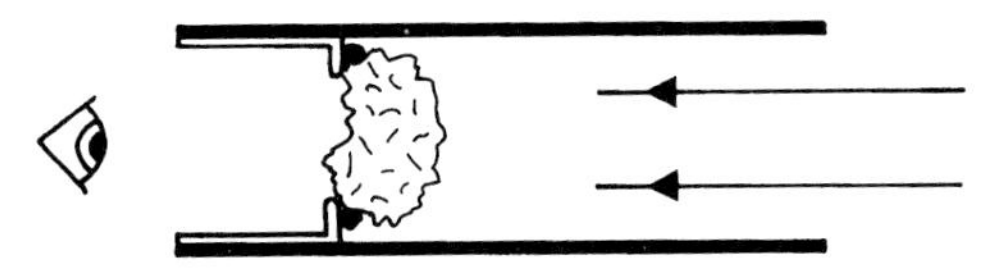

Figure 6.4. A possible sunstone—a 'natural' polariscope made from a cordierite crystal. A screw bottle top with a hole cut in it supports the rough crystal within a short cardboard tube. The direction of polarisation in the crystal can be determined with a piece of polaroid and marked on the outside of the tube. The direction of the sun can then be found by pointing the tube at the sky and watching the crystal darken as the tube is rotated. Compare this device with the 'synthetic' version shown in colour plate 7.

the crystal rather than trying to see through it. The tube is then rotated until there is a noticeable darkening within the crystal and reference marks previously made on opposite sides of the tube will then point at the sun. Two such bearings from different parts of the sky show just where the sun is. This works surprisingly well even when the sky is overcast, as long as the cloud cover is not too thick and dark. It does not work nearly as well if the crystal is smooth and polished; greatest sensitivity seems to depend on the changing brightness of glints within the rough stone. It works rather better than the synthetic polariscopes described in chapter 2, and very much better than the simpler ones described in chapter 1 or a calcite dichroscope as shown in figure 3.6.

This arrangement may also help to understand an anecdote from Gudmund's Saga. He was murdered and robbed and his sunstone was discarded on the beach as worthless, but later it was recovered. It seems most unlikely that a single rough crystal could ever be found again on a beach, but it might have been mounted in a tube, made perhaps of horn, for viewing the sky. This might have seemed of no value to an ignorant person but would have been instantly recognisable as a navigation instrument to anyone familiar with its use. This last bit is my own speculation, because such a device works for me, but I am greatly indebted to Jørgen Jensen of Skodsborg, Denmark, for telling me of these stories and for a stimulating correspondence.

Much more recently, sophisticated sky compasses were developed for transpolar aviation around the 1950s. Before satellite or inertial navigation, and when even radio beacons were scarce, polar navigation presented problems. The magnetic compass is unreliable near the poles, indeed it is useless near the magnetic poles, and there are extended periods of twilight when neither the sun nor the stars are visible. But the instruments called the Pfund and the Kollsman sky compasses were elaborate polariscopes that were used to observe the polarisation in the sky above the aircraft and so get a bearing on the sun. Apparently Kollsman sky compasses were used by SAS navigators on direct flights between Copenhagen and Anchorage. The great circle route between these cities passes quite close to the North Pole and even closer to the North Magnetic Pole.

Another application of sky polarisation is in the 'polar clocks' invented by Charles Wheatstone at King's College, London and described by William Spottiswoode in 1874. The sky around the Pole Star is viewed through a polariscope such as a Nicol prism, with some retarder crystals to make the contrast sharper. The instrument is then rotated until it indicates the direction of sky polarisation, which of course rotates around the pole star during the day at 15° per hour, being horizontal at noon and vertical at 6am and 6pm (figure 6.3). In an ingeniously simple form (figure 6.5) with no moving parts, the original polar analyser is a sheet of glass blackened behind and mounted to view the polar sky at Brewster's angle (see chapter 7). A fan of retarder crystals formed of slivers of selenite (gypsum) radiates across the view and the most strongly coloured crystal indicates the time without the need for any adjustment. Three such instruments are now displayed at the Science Museum, at the Old Greenwich Observatory and in King's College, London, and there is another at the Royal Institution. A clock of this kind can easily be improvised with modern materials, using either reflecting glass or a piece of polaroid and a fan of cellophane slips as retarders.

The advantages of a polar clock over the conventional sundial were listed by Wheatstone as: (i) it always faces the same part of the sky and so is not affected by shadows of trees, buildings or mountains—indeed it can itself be mounted in permanent shade as long as it can see the northern sky, whereas a sundial must be exposed to the full arc of the sun's path; (ii) it works for some time after sunset and before sunrise; (iii) it also works 'when the sky is overcast, if the clouds do not exceed a certain density'. This last point supports my own experience with the cordierite polariscope used as a 'sunstone'. A modern variant of the polar clock has been set up as a sculptural curiosity in Dornach, Switzerland.

Finally, the polarisation of scattered light is used in astronomy. For instance the corona around the sun is visible mainly because its

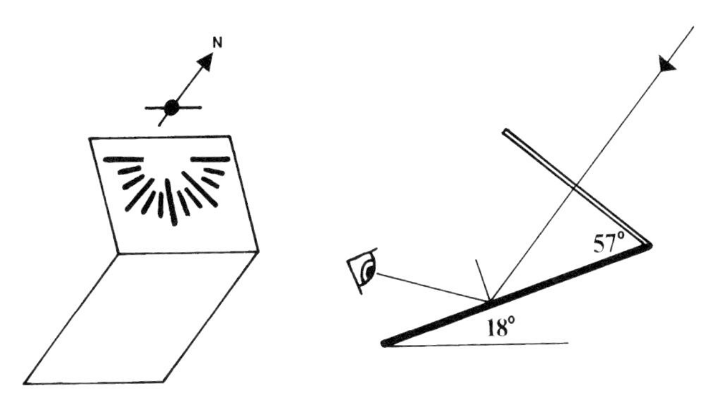

Figure 6.5. Wheatstone's polar clock. Light from the region of the sky around the Pole Star passes through a sheet of glass, parallel with the equator and bearing a radial 'fan' of selenite or mica crystals that act as retarders. It is then viewed after reflection at Brewster's angle from a sheet of glass painted black underneath. This acts as a polar analyser (see chapter 7) and the time of day is indicated by which crystal appears most strongly coloured. The angles shown are appropriate for use at the latitude of London (51°28′N).

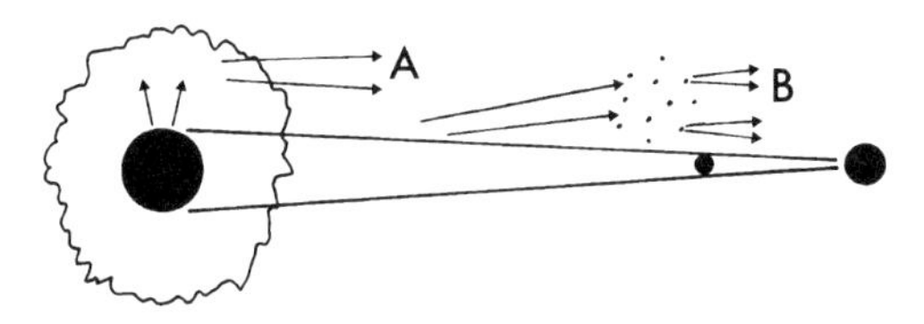

Figure 6.6. Viewing the solar corona during a total eclipse of the sun when direct sunlight is blocked by the moon. Light scattered at right angles by the corona (A) is strongly polarised. The strong obscuring glare of sunlight scattered forwards by space dust (B) is not polarised and can, in principle, be separated by the careful use of polar filters (not drawn to scale).

clouds of electrons scatter sunlight towards us (and its spectrum shows a clear inverse fourth power relation to wavelength). But sunlight coming directly towards earth is also scattered forwards by space dust (figure 6.6) and this largely obscures the corona, just as the glare around the headlights of an oncoming car in fog obscures objects behind it. However, the corona light is mainly scattered towards earth at right angles to its orignal path and is therefore strongly polarised whereas the forward scattered light is not (nor does its spectrum show the inverse fourth power relationship because the dust is much larger). During total eclipses, astronomers use polarising filters to subtract one component from the other so that a clearer picture can be obtained of the 'true' or K-corona.

Another astronomical application has been the analysis of the clouds that surround the planet Venus and completely hide its surface from our view. The polarisation of light scattered by these clouds was first studied in 1922, and in 1971 a very ingenious analysis was able to establish the following facts: most of the cloud particles are spherical and their mean diameter is close to 2 μm; their refractive index at a wavelength of 550 nm (yellow–green) is 1.45 ± 0.02 so that they cannot consist of pure water or ice; and they float high in the atmosphere where the pressure is only about 50 mbar. That is an impressive body of information to obtain from observations of polarised light by telescopes here on earth and it depends almost entirely on the polarisation produced by scattering.

Chapter 7

Reflection

One afternoon in 1808 a French scientist called Etienne-Louis Malus discovered something remarkable about light reflected from transparent materials. From his home in the rue d'Enfer in Paris, Malus examined the sunlight reflected from a window in the Palace of Luxembourg, just over 1 km away to the north-northeast. Looking through a birefringent crystal, he expected to see two images of equal brightness (see chapter 3) but instead he found that if he rotated the crystal around the line of sight, each image from the window was dimmed in turn every 90°. As night was drawing on, he continued his observations with candlelight reflected by glass (as in figure 7.1) and also when reflected in the surface of a bowl of water. He found that the effect was most marked when the angles of incidence and reflection were around 55°, though slightly more for glass than for water.

The details of this important story often vary in the telling: sometimes it is said a calcite crystal was used, sometimes a quartz, and even the date is quite often misquoted. Malus' early papers on the subject (from December 1808) do not seem to relate the original circumstances, which were anecdotally recounted only in 1855 by his friend Francois Arago, in a posthumous appreciation of Malus commissioned by the Academie des Sciences. Arago only referred to the use of 'a doubly refracting crystal', but he was also discussing calcite in detail in adjacent paragraphs. The use of quartz is usually much trickier because the two images overlap so much (see chapter 3) and therefore might seem less likely. On the other hand the glint from a window 1 km away is quite small and could give separated images through a quartz crystal, although a candle flame would probably be too large a source in a domestic room.

Figure 7.1. Polarisation by reflection in a sheet of glass. A vertical light box (a backlit translucent screen) has a piece of polaroid propped against its lower right, with the polarisation direction vertical. Light from the screen reflects well from a horizontal sheet of clear glass and also passes quite well through the polaroid. But light from the polaroid is not reflected in the glass because it is incident at around Brewster's angle at which only horizontally polarised light can be reflected.

The Palace of Luxembourg housed the Senate as it still does today. The garden, which is open to the public, now has tall plane trees and there are tall buildings near the Passage d'Enfer that would almost certainly obscure the view, even from the top floor. But one can get a splendid bird's eye view of the whole area from the top of the 200 m high Montparnasse Tower, a little more to the west and a comparable distance from the Palace. From there I found that a small quartz column easily resolves glints from the Palace as double images. A visit to the Palace garden also showed that the windows open on hinges, thus allowing the possibility of being set at a reflecting angle, although the building itself is arranged exactly north–south. But the sun must have been low that day in 1808 for its reflection to be seen at a distance over level ground, so the event must have started late in the day. In mid-summer the setting sun, as reflected to the Passage d'Enfer by a suitably opened window, would have an angle of incidence within about 7° of Brewster's angle (see p 74)—close enough to give a high degree of polarisation since the maximum effect is not at all critical.

Those seemingly simple observations have been hailed as a great turning point in our understanding of optics and the nature of light. At that time there was great puzzlement over the two rays produced from a single source by a birefringent crystal (see chapter 3). When two similar crystals are superimposed, the two rays may be split into four; but as one of the crystals is turned, the four become two and at one point (provided the two crystals are equally thick) even fuse to become one. This was thought to be a curious and inexplicable property of the crystals themselves, but Malus now showed that the nature of the light itself is different in the two rays. Light reflected from glass or water was clearly unusual in some way since it could form either of the beams normally produced by a birefringent crystal, depending on the orientation of the crystal. This light must therefore have some characteristic of its own that is expressed at right angles to its path.

This new way of producing polarised light by reflection very soon led a number of investigators to develop the first explanations both of double refraction and of the nature of polarisation itself. In 1808 the Academie des Sciences in Paris had offered a prize for a theory of double refraction in crystals and it was awarded to Malus in 1810; he died in 1812 aged only 37. Polarisation by reflection has also proved to be of great practical importance because in all optical devices every surface, including those of lenses, introduces some reflection and therefore some polarisation, and this may affect their operation, as discussed in detail later.

We now know that the two beams from a birefringent crystal are polarised in opposite ways and that Malus's observation of alternate dimming of the beams shows that light reflected by shiny, non-metallic surfaces is itself polarised. The direction of such polarisation is at right angles to the plane of the incident and reflected rays and this gives a simple way (as promised in chapter 1) to find which way any given piece of polaroid is aligned. Look through it at a reflection from any horizontal, shiny surface (water, glass, polished wood or gloss paint) and rotate the polaroid until the reflection dims. The direction of polarisation for transmission through the polaroid is then vertical and can be marked in one corner.

Malus soon showed that any light reflected from a piece of glass at an incidence of 57° can be reflected again from another, parallel piece (thus also at 57°), but when the second piece is rotated around the incident axis (figure 7.2) its reflection is extinguished twice in every turn. At angles of incidence other than 57° the dimming is less effective.

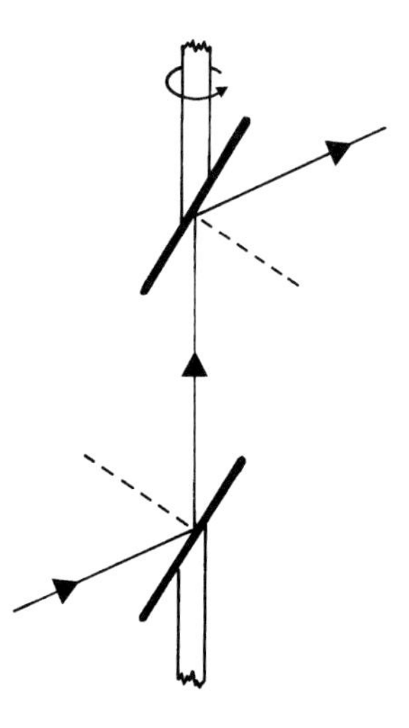

Figure 7.2. Malus's experiment. Two parallel glass sheets reflect light at an angle of incidence of 57°. But when the upper one is rotated around the vertical axis, the emergent beam fades twice as it swings round in a circle. The plates act as polariser and analyser, or second polariser.

The same effects were seen when light was reflected from glass and water in turn. Then in 1814 David Brewster realised that this 'best' angle of incidence is the one whose tangent is equal to the refractive index of the reflecting material. This is now known as Brewster's angle, at which reflected light is fully polarised. Thus the Brewster angle for glass is 57° because its tangent is 1.54 and that is the refractive index for glass. Brewster's angle also has the property that the reflected ray and the refracted ray entering the material are at right angles to each other (figure 7.3).

At Brewster's angle, only around 15% of the incident light is reflected but as it is very nearly 100% polarised, a simple reflector can be used as a very cheap polariser of wide aperture. As Malus found, a second reflector can act as a second polariser or polar analyser, reflecting the light to a varying degree, depending on its orientation to the direction of polarisation from the first reflector. This then completes a simple but very effective home-made polarising apparatus. The reflectors are best made from black perspex (plexiglass) but clear perspex will do perfectly well if the back face is painted black (without the paint there

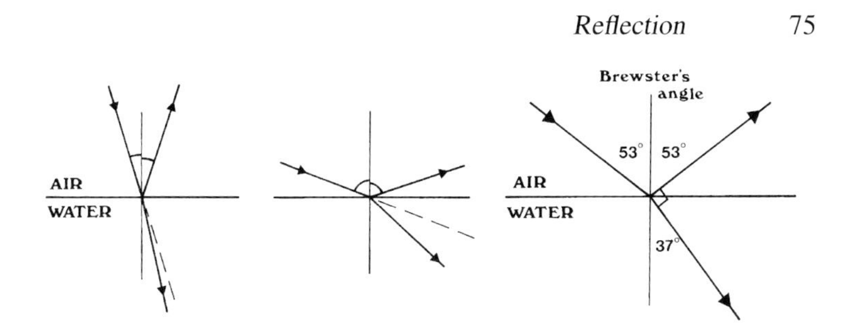

Figure 7.3. Brewster's angle. At the surface of a transparent material a beam of light is partly reflected and partly refracted. At high incidence (left) the two ongoing beams form an obtuse angle and at low incidence (centre) an acute angle. They are exactly at right angles to each other (right) when the tangent of the angle of incidence is equal to the refractive index (57° for glass in air). The reflected light is then virtually 100% polarised, in a direction normal to the plane of the diagram.

are distractions due to both transparency and double reflection). Glass is even cheaper but it is fragile. The plates, say about 10 cm × 15 cm, must be mounted in rectangular cardboard frames so that they lean across the axis at 34° (i.e. $90° - 56°$ for perspex, or $33° = 90° - 57°$ for glass, to be rather unnecessarily precise). A hole cut in the cardboard wall completes the optical pathway (figure 7.4). A sheet of clear perspex or glass, placed on top of one frame, can then support the other and also the objects to be examined. Relative rotation around the vertical axis allows the polarisers to be set parallel or crossed as required. With a little ingenuity the whole apparatus can be made to fold flat when the reflectors are removed so that it can all be stored in a shallow box.

This very simple device, or variants of it, can be used to demonstrate most of the vivid effects described in chapters 2 and 3 without the expense of large sheets of polaroid. Indeed until the 1930s reflection from glass was the only way to obtain large polarisers since both Nicol prisms and tourmaline crystals were necessarily small. A pair combined as shown in figure 7.4 was called a reflecting polariscope. It does have two drawbacks, however. First, the effectiveness is somewhat degraded towards the edges because the incident light and the line of view cannot be strictly at Brewster's angle across the whole field, especially if source and viewer are nearby. Second, the reflection is rather weak (at most

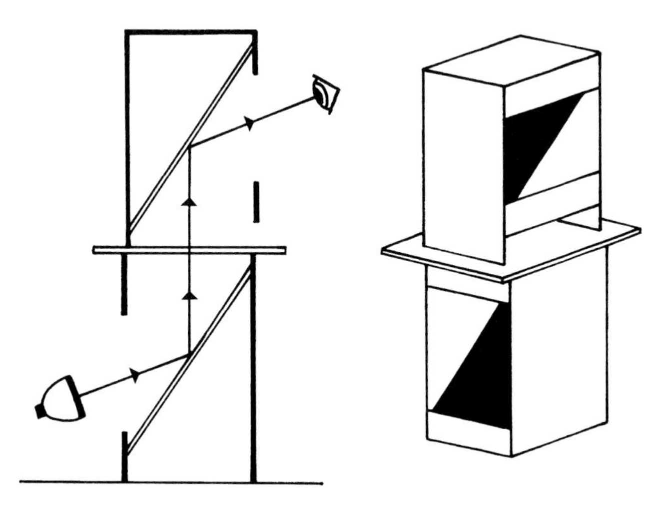

Figure 7.4. A reflecting polariscope made with two perspex sheets in cardboard stands. Left: a section showing how light is reflected twice at Brewster's angle giving the bright field effect of parallel polarisers. Right: a view of the device with the top rotated to give the dark field effect of crossed polarisers.

only about 15% of the incident light is reflected at the first surface) but this can easily be compensated for by using a bright source such as a small halogen lamp.

One way in which the brightness can be increased is to use a stack of parallel, transparent plates and take either the reflected or the transmitted light, usually the latter. At Brewster's angle the light reflected from one surface is dim but very strongly polarised. So the remaining light that passes onwards through the plate will be partially polarised by the subtraction. This light then meets the second face of the plate at Brewster's angle (due to refraction the angle is now different from the original incidence but it matches Brewster's angle exactly since in a denser medium it is the cotangent, not the tangent, that is equal to the refractive index). Light reflected back into the plate, therefore, is again totally polarised and the light passing on into air is a little more polarised (figure 7.5). So it would seem that passing this light through other, parallel transparent plates will reflect away more and more of the light polarised in one direction, making the remainder progressively more polarised. If the plates are very clear, this ongoing beam may be very

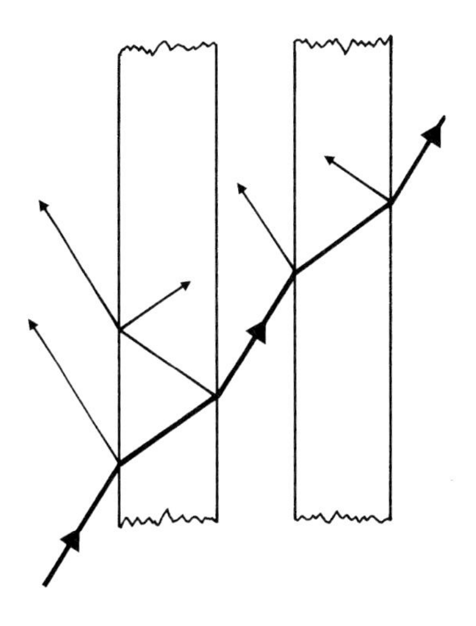

Figure 7.5. Multiple reflections by more than one sheet of glass. Light reflected from the first face at Brewster's angle is not bright but it is completely polarised; light refracted from this face meets the second face at Brewster's angle for the denser medium and the reflection is again fully polarised. The ongoing light (heavy line) is therefore enriched in light polarised at right angles—within the plane of the diagram. A second and further plates repeat the process so that the emergent beam is bright and, in principle, becomes progressively more polarised. But all the reflected light does not necessarily leave the system because some of it is reflected again as shown by the finer lines, and some of this 'unwanted light' can re-enter the onward path to degrade the final degree of polarisation.

bright—approaching 50% of the incident intensity. Many textbooks say a high degree of polarisation is quickly achieved, by as few as eight plates, and calculations of simple repeated subtraction would seem to support this.

But in practice things are rather more complicated and less effective. The main problem is that if the successive plates are close together, the light reflected from each face does not leave the system but is reflected again by the adjacent plate. It will also be reflected internally within each plate. Both processes may be repeated many times for each ray and can only be avoided by quite elaborate design. Proper analysis shows that even 16 plates in a simple stack at Brewster's angle can only achieve 74% polarisation. This theoretical performance is greatly degraded by dust or by imperfections on the glass surfaces. Adding more plates gives diminishing returns and begins to add serious loss due to absorption of the transmitted rays. Indeed a simple stack of thin glass plates does not even give the greatest polarisation at Brewster's angle but at a much higher, near-grazing angle of incidence, as noted by Brewster himself in 1831. For most purposes, therefore, a plain stack of plates is not a practical solution despite its apparent simplicity. Nevertheless a sophisticated modern version can be made by building a layer of say 25 very thin films between the faces of two prisms; by adjusting the thickness of the films, all less than a wavelength, it is possible to achieve a high degree of polarisation up to 98%, at least over a narrow range of wavelengths.

Iridescent natural materials, such as some butterfly and beetle colours, some fish scales and feathers, and the mother of pearl in shells, are composed of multiple microscopic layers of material whose very many internal reflections interfere to produce different colours in different directions. The gemstone opal gives similar effects from a structure composed of an array of microscopic spheres of silica. As might be expected, the reflected coloured lights from these materials are sometimes quite strongly polarised, as can be seen when they are examined through a rotated polaroid. Most of the bright glints of hoar frost sparkling in sunshine are also strongly polarised in a direction depending on the angle of view; the few exceptions are presumably reflecting well away from Brewster's angle and so remain bright when seen through a rotating polar.

Malus himself found that reflections from the surface of water are strongly polarised in just the same way as for shiny solids, with a Brewster angle of 53°. This accounts for the usefulness of polaroid sunglasses. It is seldom that the actual brightness of light is distressing to the eye but rather the glare: unwanted light coming from below. Much of this is horizontally polarised, especially when reflected from wet roads but also quite strongly from most kinds of dry ground. Polaroid sunglasses are vertically polarised and so they greatly reduce glare even when the line of sight is not very near Brewster's angle. (The downward looking parts of some insect eyes are sensitive only to vertically polarised light, presumably to achieve the same end when flying.) Polaroid glasses give a quite dramatic effect when one looks into still water, for reflections at the surface are reduced and at Brewster's angle are effectively abolished. One can then see clearly into the depths, with details of the bottom, weeds and fishes that are normally obscured by reflection of the sky or clouds. The water surface may seem to disappear

Figure 7.6. A sound in Bermuda photographed through a vertically orientated polaroid filter. The bather is sitting near the edge of the water which is very clear and quite still. When the polaroid suppresses surface reflections the water virtually disappears.

completely (figure 7.6) provided that there are no ripples.

The polarisation of light reflected from water is admirably exploited by the aquatic bug called the Water Boatman or Backswimmer (*Notonecta*). As its names suggest, this insect lies on its back at the surface of freshwater and rows itself around by means of greatly enlarged back legs (figure 7.7). But from time to time it flies off looking for a new pond or ditch. In 1935 it was noticed that just before plunging in, when less than a metre above the surface, the insect hesitates briefly and tilts itself upwards by nearly 20°. This was explained only in 1984 by careful examination of the insect's vision. The ventral facets of its compound eye each contain eight sensory cells of which two are sensitive to ultraviolet, one polarised vertically, the other horizontally. When swimming, these facets look up at the sky and would seem to be ideal polar analysers for sky compass orientation (chapter 6). But no facets of this kind look forwards—at all angles more than 35° from vertically below (figure 7.7) all the polarisation sensitive cells respond only to horizontal polarisation. In flight, these forward-looking facets

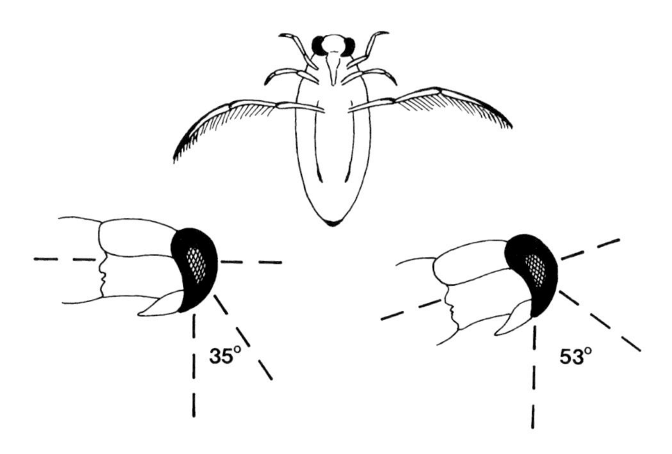

Figure 7.7. The Water boatman (*Notonecta*): top, swimming on its back; below, side views of the head; left, in normal flight posture; and right, tilted up by 28° before plunging into water.

would be ideal for detecting the polarised light reflected from standing water because light that is horizontally polarised is more likely to be seen if all the visual cells are maximally sensitive to light polarised in that direction, although they would not be able to tell that it is, in fact, polarised. But during the final plunge, the upward tilt brings some downward-looking, analysing facets just up to Brewster's angle. Maximum stimulation of some cells and loss of stimulus to the others would then confirm that the light does come from reflection off a liquid and the bug can safely complete its dive. In ingenious laboratory tests, bugs landed on upward-facing lamps but only if they emitted ultraviolet light that was polarised transversely to the line of approach, thus simulating light reflected from water.

Dragonflies also have downward-pointing facets able to analyse the polarisation of either ultraviolet or of blue light. As before, these probably serve to detect the presence of water. It has also been suggested that, as the dragonflies fly over an expanse of water, the reflected light from all sides is predominantly horizontally polarised and could provide an 'artificial horizon' to help in maintaining a proper flight path. Such cues can sometimes be misleading, however, for recently it has been found that some mayflies are attracted in very large numbers to mate and lay their eggs on asphalt road surfaces, which is, of course, disastrous

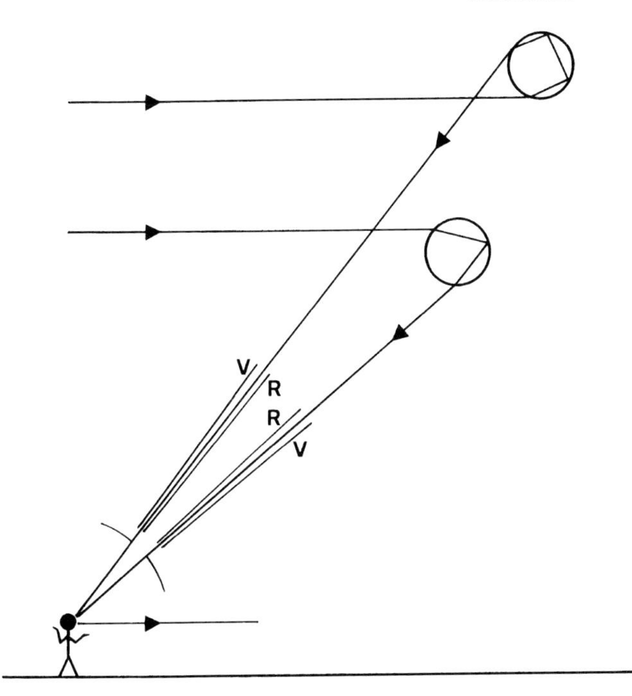

Figure 7.8. Two ray paths through raindrops that give rise to rainbows. The lower case involves a single reflection and gives the primary bow at 42° to the sun–antisun axis while the upper case involves two reflections and produces the secondary bow further out. In both cases the coloured rainbows are only chromatic fringes of larger areas of white reflective scattering which is strongly polarised in the tangential directions, around the bows.

for reproduction. It seems that roads not only resemble streams in their general shape, but even their reflected light is polarised to about the same degree as that from rippled water, which polarises light less effectively than a smooth surface.

Rainbows are another familiar source of strongly polarised light. It is often thought that rainbows result from the backscattering of sunlight by rain droplets, but actually the light is mainly returned by reflection within each drop. Figure 7.8 shows how light entering a spherical drop of water is first bent by refraction, then reflected from the back of the drop and finally refracted again as it leaves. Light can only be returned in this way up to an angle of about 42°, resulting in a disc of reflected light whose edge is 42° from the anti-sun point. This is seldom actually noticed although it shows clearly when viewed against a dark background (colour plate 25). The actual familiar rainbow is just a coloured fringe at the outer edge of this disc. It occurs because refraction allows the longer red waves to be returned at a slightly greater angle than the shorter blue and violet waves, thus forming a narrow spectrum, about 2° wide.

About 8° further out, light is again returned after being reflected, this time twice within each water drop. The inner edge of this band forms the secondary rainbow, about 3° wide and with the colours reversed—red on the inner side and blue–violet outwards. In both rainbows, the angles of reflection inside the drops happen to be close to Brewster's angle (37° or the cotangent of the refractive index for light within the denser medium) so that both kinds of rainbow are strongly polarised, the primary bow slightly more so (96%) than the secondary bow (90%). The direction of polarisation is tangential to the bows themselves, that is it runs around their curves, and this effect can easily be seen with a piece of polaroid. Segments of the bows themselves and the white regions bounded by them can be made to disappear when the polaroid is orientated in the radial direction.

When the sun is higher in the sky than 42°, primary bows can seldom be seen because they would have to be below the horizon, but bows do appear if the droplets are not falling rain but dew drops on a lawn, or the spray from a garden sprinkler or from a waterfall (colour plate 25). These rainbows are easier to observe than those produced by rain showers because they are generally less transitory and do not fade just as one is getting interested in making observations! There are many other atmospheric phenomena that give rise to colours in the sky and these are often polarised too. They are called haloes and arise from various kinds of ice crystals in high clouds; they are described in detail in the book by Greenler listed in the bibliography. Curious things happen to the polarisation of rainbows that are reflected by standing water and these are discussed in the book by Konnen, also listed there. Unfortunately both topics are too complex to go into here.

Photographers often use linear polarising filters on their cameras, either to suppress reflections or to obtain striking contrasts between sky and clouds (chapter 6). But they are often puzzled when catalogues offer filters of 'circular polaroid'. The reason is that many modern cameras have reflectors behind the lens that divert some light for automatic focusing or for automatic exposure control. These reflectors polarise

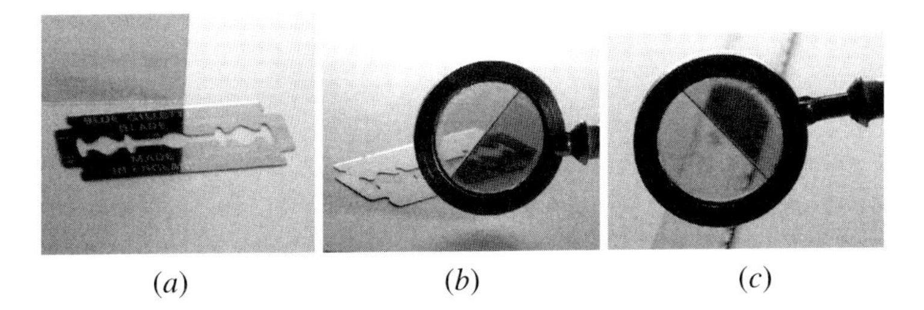

(*a*) (*b*) (*c*)

Figure 7.9. Light reflected from dark metals such as blue steel may be strongly polarised, especially at near grazing incidence—as if the refractive index were very high. (*a*) The left half, being lit here by vertically polarised light, is dark and unreflective. Other metals may show some degree of polarisation unless they are very shiny and reflect nearly all the incident light. Polarised reflection is also shown for a razor blade (*b*) and a blue hack-saw blade (*c*) using the polariscope of figure 1.4.

this light at least partially, so that a linear polariser on the lens will affect the readings as it is rotated. To avoid this, the back of the filter is covered by a quarter-wave retarder that changes the emerging light from plane to circular polarisation (see chapter 8). Because the light is now radially symmetrical, rotating this filter makes no difference to any reflections and automatic readings within the camera. The circular polaroid is actually fitted 'backwards', not in the way that would be needed to respond, as the name suggests, to circular polarisation of the incoming light. This is just one example of the problems that can arise by reflection within optical instruments; it is an effect that can easily give rise to false readings in a wide range of instruments unless proper precautions are taken in the design.

It is frequently stated that light reflected from metals is not polarised but once again reality is not quite so simple. Light reflected from dark metals such as tempered 'blue' steel may be very strongly polarised, especially at high angles of incidence (figure 7.9). Aluminium, copper and stainless steel also give quite strong polarised reflections at grazing angle (figure 7.10). It has also been long known that light that is emitted by white hot metals, either solid or molten, is strongly polarised when viewed at highly oblique angles. It was the absence of such polarisation in light from the 'edges' of the sun that showed it must be made of gas rather than solids. On the other hand a degree of polarisation in moonlight confirms that it shines by reflecting light from the sun.

Figure 7.10. Light reflected by polished copper (bottom) is much more strongly polarised than light reflected from brass (top) which is brighter and more reflective overall. Both here and in figure 7.9, the polariscope has two pieces of polaroid each orientated at 45° to the line where they meet, so the reflected light is shown to be horizontally polarised.

Bright shiny metals reflect a very high proportion of the total incident light, so clearly this cannot be strongly polarised although there is a more subtle effect. Light vibrating in the direction normal to the surface (which would be suppressed by the Brewster effect with non-metals) suffers a phase change with respect to any components vibrating at right angles. So if the incident light is already polarised, it may become circularly or elliptically polarised after reflection (again see chapter 8 for further details). This effect will not be evident when examined with linear polaroid and is therefore often overlooked. Light reflected by a back-silvered glass mirror shows little effect because refraction by the glass limits the effective angle of incidence: even a grazing angle at the surface of the glass is reduced to a much lower angle of incidence on the silvered back (maximum about 41°, see figure 8.11).

A very interesting case is a crystal of germanium, a semiconductor that polarises light like a shiny non-conductor. Its refractive index has

Figure 7.11. Reflection on a piece of glass as in figure 7.1 but with pictures laid beneath the glass. Here it is seen that where the light is vertically polarised, and so is not reflected, the glass becomes 'transparent' and objects below it are no longer obscured by reflections. Such an effect might help fishing birds to see through the surface of water.

the extraordinarily high value of 4 so that Brewster's angle is 76° and complete polarisation occurs at an almost grazing angle. But the degree of reflection is about 40% which is almost metal-like since a polarised beam obviously cannot exceed 50% of the original unpolarised light.

The effect of polarisation can sometimes be seen without the aid of an artificial analyser. Colour plate 26 shows a lake called Bachalpsee above Grindelwald, Switzerland, photographed at dawn. The distinctive peaks of Schreckhorn and Finsteraarhorn show the view to the south-southeast (the latter bearing about 155°) and the shadows show the summer sun must be rising in the northeast over to the left. The striking feature is that the nearer part of the lake does not reflect the brilliant blue sky. The reason is that the clear sky 90° from the sun must be very strongly polarised vertically (see chapter 6); but the nearer water gives a line of view that approaches Brewster's angle at which only horizontally polarised light can be reflected. So the lake is acting as an analyser 'crossed' with the sky's polariser.

Another example is seen in colour plate 27, looking due north at sunset, as shown by shadows on the dome and minarets. The nearer part of the reflecting pool does not reflect the clear sky although the Taj Mahal itself is reflected perfectly because it is not a polariser. This beautiful effect is often seen in photographs where the view combines light from the side, a clear sky and still water in the near foreground. But are there any depictions of the effect in art? I would be most grateful for any information helping to identify a good example of the effect being noted by a landscape painter.

Finally one might speculate about the implications of polarised reflections for fishing birds. If, as now seems probable, some birds are able to respond selectively to the plane of polarisation, they might gain the advantages of polaroid glasses as modelled in figure 7.11. It would be interesting to see whether herons and kingfishers, for example, attack their prey at Brewster's angle of 53° where they would be able to see best into the water. Even if they do not respond directly to polarisation itself, they would still benefit by exploiting Brewster's angle at right angles to the sun when the sky is blue, as we can ourselves. The advantage would be especially great in early morning and late afternoon if birds faced north or south rather than in other directions. Simple observations of the behaviour of such birds might be very useful and interesting.

Chapter 8

Going circular

So far only linearly polarised light has been considered. It occurs very commonly in both science and nature but it is actually a special case of a more general system which also includes circular and elliptical polarisation. Examples of these are much rarer but are of considerable interest anyway. In most cases the circular polarisation is derived from linearly polarised light which is itself very common. The theory may seem a little abstruse at first, but it is worth following in order to understand the larger picture.

The explanation of 'changing direction' given in chapter 2 was rather oversimplified because it was deliberately limited to the case of retardation by half a wave (the half-wave plate). Although thicker and thinner retarders were then discussed briefly, the consideration of vectors in such cases was felt to be an unnecessary distraction at that point. The half-wave plate is simple to understand because the two emergent vectors always recombine to produce another linear vector and, as a result, linear polarisation is preserved on emergence, though the direction of polarisation may be changed. However, if a retardation of only a quarter of a wave occurs, the emerging components combine to form light in which the vector that represents the direction of electrical oscillation actually rotates through one complete revolution during each wave. This is called circular polarisation. It has no single 'direction' of polarisation but the vector can rotate either clockwise or anticlockwise (as seen from behind—i.e. from the source) and is called right circular or left circular polarisation respectively.

If all this seems a bit bizarre (and such a wave is certainly difficult to imagine at first), consider the view of a helical spring shown in

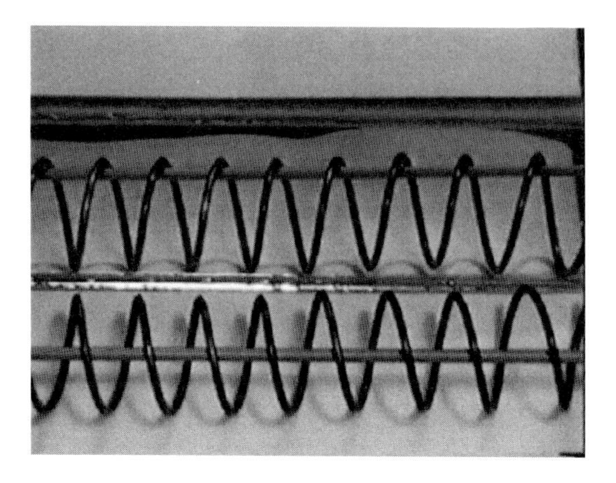

Figure 8.1. A coiled wire spring suspended from a straight wire and photographed from above (lower image) and also seen simultaneously from the side in a mirror inclined at 45° (upper image). Both profiles look just like a simple, sinusoidal wave but they are staggered by just a quarter of a wavelength as shown by the points of intersection with the straight wire—at the midpoint below and at the peak in the reflection.

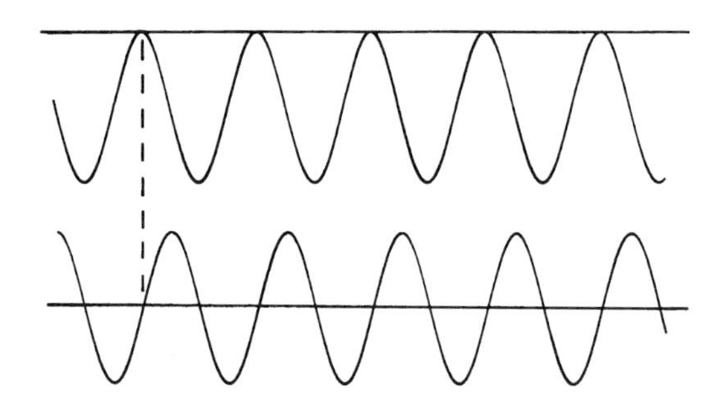

Figure 8.2. A diagrammatic reconstruction of figure 8.1. Two simple sine waves are a quarter of a wave apart and the dotted guide line shows the relative alignment of the two waves.

figure 8.1. From one side, the helix looks just like a simple wave—in fact if perspective is overlooked (i.e. if the spring is viewed from a great distance as drawn in figure 8.2) the wire is exactly the shape of a sine wave. The same spring viewed at right angles to the first view, as seen in the inclined mirror in the figure, is still a sinusoidal wave but the 'two'

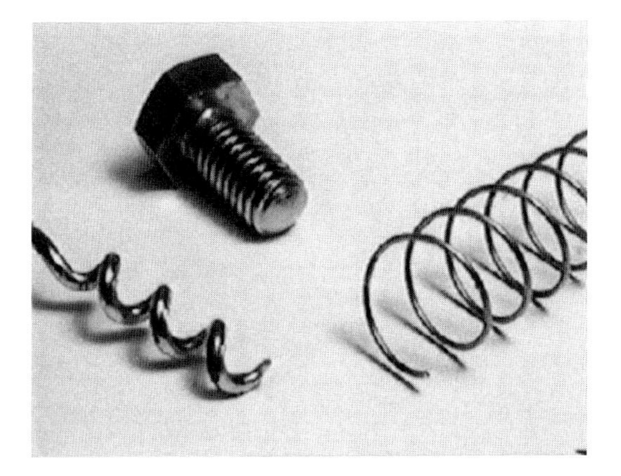

Figure 8.3. Some familiar objects with the form of a right-handed helix. When reflected in a mirror, all would appear to be coiled in a left-handed manner.

waves are just a quarter of a wave 'out of step'—the peaks and troughs of one view are the nearest and furthest points in the other view. One knows that in three dimensions the wire traces round a circle, once for each 'wave' that is seen from the side. Other familiar examples of the same shape include the thread of a screw and a corkscrew (figure 8.3). In the former, the thread generally rotates clockwise as one traces it away from the point of view and it is then called a right-handed thread; in the less common 'left-handed screws', the threads rotate anticlockwise away from the viewpoint; left-handed corkscrews are also occasionally seen, being made to help left-handed people. So the description of circular polarisation with a rotating vector as given here also applies to some very familiar objects. It is rather like the schoolboy challenge to 'describe a spiral [sic] staircase without using your hands'—it always seems unnecessarily complicated when put into words alone.

The fact that two waves vibrating at right angles and one quarter of a wave out of step will generate a circular motion for every wave is also seen in two simple models. If a pendulum is suspended from another pendulum that swings at exactly the same rate but at right angles to its own swing (figure 8.4), the path of the lower bob will depend on the relative timing of the swings. If the two are in step because each was started at the end of its swing at the same moment, then the bob moves in a straight line—say from back left to front right and back again. Starting one of the pendulums at the other extreme, just half a complete

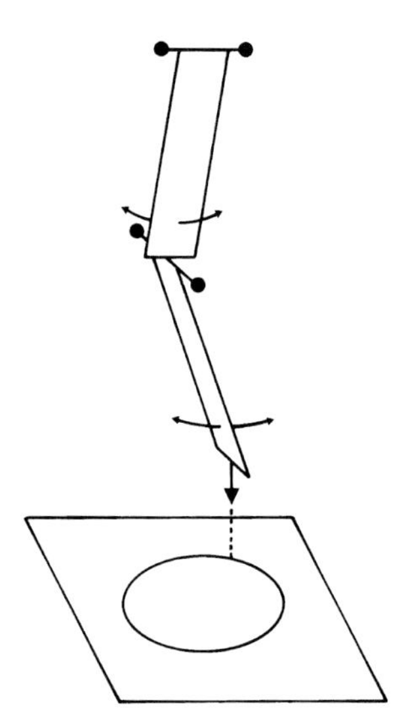

Figure 8.4. One pendulum suspended from another pendulum. The two swing with the same period but in directions at right angles to each other. Changing the relative timing (phase) of the swings produces a series of different patterns of movement that can be demonstrated by letting sand trickle from the lower bob. When the two swings differ by a quarter of a wave, the sand is deposited around a circular path.

swing later, makes the bob swing in a straight line at right angles to the first path. This change in relative timing by half a wave is exactly analogous to the half-wave plate that retards one wave and so twists the resultant vector through a right angle (as seen in chapter 2). But if one pendulum is delayed by a quarter of its complete swing, then the lower bob swings round in a circle. Imagine it swinging over an upturned clock face. Starting from far left and middle distance (at 9), it moves nearer and to the right; at the nearest point it is only half way to the right (at 6) and continues to the right as it recedes (to 3); starting back to the left, it reaches its furthest point (at 12), and finally completes its leftward swing as it comes forward again (to 9).

The second illustrative model is an electronic version of the same thing. The glowing spot on the screen of an oscilloscope screen can be made to swing left and right by applying a sinusoidal deflecting signal. Another such signal can be applied to make the spot move up and down

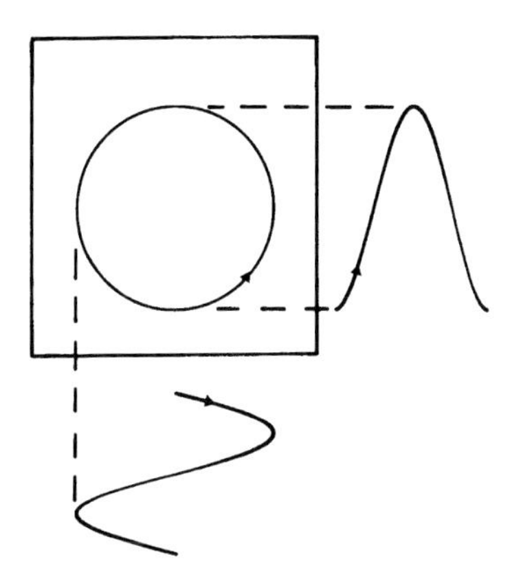

Figure 8.5. The bright spot on an oscilloscope screen can be deflected from side to side and also vertically up and down by application of sinusoidal waves. When the two waves are identical but staggered by a quarter of a wavelength, then the spot revolves in a circle. This is a special case of a Lissajou's figure.

on the screen. When the two waves are in step, the spot moves in a straight line diagonally; when they are a quarter of a wave out of step the spot moves in a circle (figure 8.5). These models can be used to investigate other situations. A perfect circle is only obtained if the two waves are exactly a quarter-wave apart and also of exactly the same size (amplitude), otherwise the circle becomes an ellipse. In the case of light we need not consider unequal waves but it is clear that 'exactly in step' and 'exactly a quarter of a wave apart' are simply two special cases of a whole range of possible timings. Starting with the two waves in step and delaying one slightly makes the straight line 'open out' into a thin ellipse (figure 8.6). Increasing the delay makes the ellipse fatter until at one-quarter of a wave difference it becomes a circle; then it gets thinner 'the other way' until at half a wave difference it becomes a straight line again but at right angles to the original one. These patterns are known as Lissajou's figures for equal frequencies, and they can be used to make very accurate comparisons between two waves.

The point of all this for polarised light is that a birefringent material resolves a wave into two orthogonal components and then retards one relative to the other. The result depends on the thickness of the material and the relative timing of the two emergent components when they recombine. The general case is elliptical polarisation for which circular

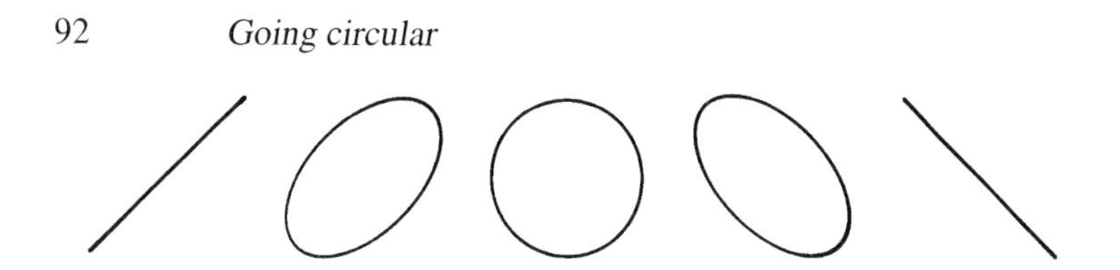

Figure 8.6. When the relative timing of the two waves of figures 8.4 or 8.5 is varied, a family of Lissajou's figures is produced. The figures shown here are the family produced by equal (1:1) frequencies and equal amplitudes; other families of figures are produced when the two waves are of different but simply related frequencies such as 1:2, 2:3 etc, but those patterns do not concern us here.

and linear polarisation are both special, extreme cases.

It is important to emphasise that elliptical polarisation is not the same as partial polarisation, although in both cases a rotating linear polar will show a partial dimming with no complete extinction of the transmitted light. Partial polarisation consists of polarised light mixed in with some unpolarised light, half of which always passes through the analysing polariser. With 100% elliptically polarised light, there is a larger electrical field vector along the long axis of the ellipse and a smaller one along its short axis. With circularly polarised light, of course, the vector is the same in all directions and there is no dimming as an analysing polar is rotated—although in practice there are usually slight colour changes due to retardations not being quite equal for all wavelengths.

Circular polarisers are therefore made by applying a quarter-wave retarder film or crystal to one side of a linear polariser, with its two special optical directions (properly called the privileged directions) at 45° to the direction of polarisation. Quite effective circular polarisers can be improvised by selecting quarter-wave cellophane films (see chapter 2) and combining each with a piece of standard, linear polaroid. Rotating the retarder by 90° interchanges the two special directions and turns a clockwise circular polariser into an anticlockwise circular polariser or *vice versa*, but in commercial materials the two films are usually bonded together. For some reason, nearly all the circular polaroid that is easily available is left-handed although the right-handed form can be obtained if needed.

A simple, if rather clumsy, method of changing the handedness or sense of rotation is to add a half-wave plate or film. This is because, if correctly orientated, retarders have a cumulative effect. A quarter-wave plus a half-wave retardation gives a three-quarter retardation, equivalent to minus a quarter; and conversely, if the extra plate is turned by 90°, a quarter minus a half also gives minus a quarter.

A circular polariser only works in one direction, with the light passing first through the linear polaroid and then through the quarter-wave plate. When light passes in the opposite direction, the random, unpolarised light is essentially unchanged by the quarter-wave plate (it is different but still random) and then emerges with linear polarisation from the polaroid. A test piece of linear polariser will immediately show the difference—on rotation it goes black in two orientions as it extinguishes light transmission on the linearly polarised side but it shows (almost) no change on the circularly polarised side.

It is not easy to distinguish between left-handed and right-handed polarisers unless one has a reference piece of known handedness. When two oppositely handed polarisers are held 'face to face' the transmission is extinguished (at all points of relative rotation) whereas two similarly handed polarisers transmit freely (even when one is rotated). A simple circular polariscope, called a Cotton polariscope after its originator, can be made from one polariser of each handedness mounted side by side. One will go dark when circularly polarised light of the opposite handedness is examined (figure 8.7). Simply by reversing this device, so that light passes the other way (first through the linear polaroids and then through the retarder films), makes it into an equally effective linear polariscope (as seen in chapter 1) since the retardation then has no effect on the visible result.

Circular polaroid films of this kind find a very useful application in greatly reducing troublesome reflections from aircraft instrument panels and radar screens. The radar screen has an image that glows by phosphorescence and instruments with dials can be illuminated from within. But in both cases lamps and illuminated objects in the room or flight deck will be reflected in the glass screens and will degrade visibility or at least create distraction. If a sheet of circular polaroid is placed over the panel, light from the display passes first through the quarter-wave retarder and then through the linear polariser, suffering only a 50% loss of intensity which can easily be compensated at source. But light from the room is circularly polarised by the filter and when it is reflected by glass the direction of rotation is reversed: left-handed circular becomes right-handed circular. This cannot pass out through the polariser because the quarter-wave retarder turns the circular into linear polarisation with its direction at right angles to the original, and this is

Figure 8.7. A Cotton circular polariscope for detecting whether light is circularly polarised and its handedness, here seen over a circular polaroid background film. Two pieces of circular polaroid, one left-handed and the other right-handed, are simply mounted next to each other. The device must be viewed from the side bearing the linear polaroids, otherwise it acts as a linear polariscope (see figure 1.4). If circular polaroid of only one handedness is available, half of it can be covered with a suitably orientated half-wave film to change its handedness. Another alternative is to mount two pieces of linear polaroid, with their directions at right angles, over a single, properly orientated quarter-wave retarder film.

then blocked by the linear polariser. One can imagine, by analogy, that the mirror image of a corkscrew, reversed in its sense of rotation, cannot be pushed through holes previously made by the real corkscrew. So a sheet of circular polaroid, when inserted the right way round, provides a 'black screen' through which self-luminous displays are seen clearly. Such components are sometimes supplied with oscilloscopes too.

A second application for circularly polarising filters is in photography for the removal of intrusive reflections from a scene or to enhance the contrast between clouds and blue sky. As described in chapter 7, the use of a linear polarising filter achieves the wanted effect but might influence the amount of light reflected by mirrors within the camera, which sample the light for metering or for automatic focusing. But if the linear polariser, having done its job of selecting one direction of vibration, is then followed by a quarter-wave retarder, the light becomes circularly polarised and the filter can be rotated to any desired position to enhance the image without affecting the strength of internal reflections. The circular polariser is therefore not actually sensitive to circular polarisation in the scene itself, as its name might suggest, since it is mounted the 'wrong way round'.

Another way to make circularly polarised light that, in some ways, is both simpler and better than using a retarder is by the use of total internal reflection. Light that is incident at about 45° upon a glass–air surface from within the glass itself is completely reflected (in contrast to external reflection where a glass surface reflects only a small part of the light). The principle is used in prisms to make good mirrors, such as those that invert the image in binoculars, where four reflections at 45° are needed and silvered glass mirrors would degrade the image intolerably. When the light is also linearly polarised at 45° both to the surface and to the plane containing the rays, it is divided into two components, one parallel with the surface and one normal to it (in the plane of the rays). These are reflected with a relative difference in timing of about one-eighth of a wave, giving elliptical polarisation. A second reflection, therefore, can give another eighth of a wave difference to make a quarter of a wave and so produce circular polarisation.

To be precise, the angle of incidence within the glass should either be about 48.5° or about 54.5°, but the former is so close to 45° that the very convenient combination of two right-angle prisms gives a broadly elliptical polarisation that is a good approximation to truly circular polarisation (figure 8.8). A prism specially designed to give two reflections at the proper angle is called the Fresnel rhomb (figure 8.9) and this gives truly circular polarisation, usually using the larger angle of incidence and reflection. Although these arrangements are not such a handy shape as the quarter-wave plate device, they do not suffer from the slight wavelength dependence that accompanies birefringence and so they avoid spurious colour effects. But they are degraded if the light does not approach from exactly the 'proper' direction and so they have a narrow angle of effective view.

Incidentally, circular polarisation is not produced within a pair of binoculars because the prisms happen to be so arranged that the effect of the first pair of reflections is cancelled by the second pair; they do not even add to give a half-wave difference that would rotate the plane between crossed polars. In addition, these considerations do not apply to the stack of plates reflecting polarisers of chapter 7, where the linear polarisation is never slanted at 45°.

Reflection is also responsible for perhaps the commonest natural occurrence of elliptical polarisation: reflection of linearly polarised light from shiny metals. In chapter 7, reflection was considered from materials

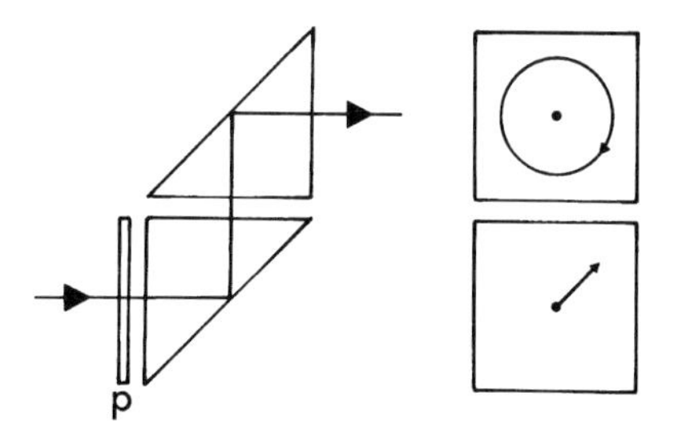

Figure 8.8. Circular polarisation can be produced by the combination of a linear polaroid and two reflecting prisms. With right angle prisms the circularity is not precise but is close enough for many purposes. On the left an oblique linear polariser (p) and two right-angle prisms; on the right the view looking into the upper prism with oblique linear polarised light converted into almost circularly polarised light. Turning the linear polariser by 90° would reverse the rotation.

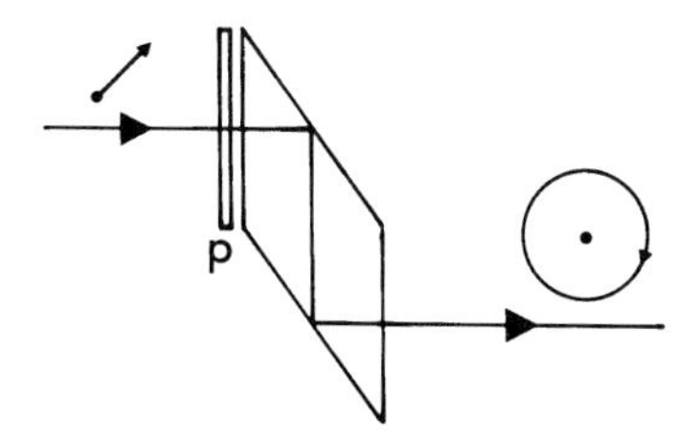

Figure 8.9. Exactly circular polarisation can be produced from oblique linearly polarised light by a Fresnel rhomb on which the angles of internal reflection are around 54° for various types of glass. Both the necessary reflections can be arranged within a single piece of glass and the circularly polarised light emerges on a path parallel to the original beam. With the linear polaroid (p) set obliquely as shown, the circular polarisation will be reversed.

that are essentially non-conductors of electricity and from dark coloured metals. It might be supposed that reflection from nice shiny, electrically conductive metal surfaces would be much simpler and it is certainly true that these reflect a much greater proportion of the incident light. But the interaction of the electromagnetic fields of light waves with such conducting materials, which have free charged electrons, is quite complex. Once again incident light is divided into two components, one polarised parallel with the surface and the other normal to it, in the plane of the light rays. As with internal reflections inside non-metals such as glass, these two components are given a timing difference on reflection from the surface of shiny metals. If the incident light is already linearly polarised at rather more than 45° to the surface, then the resultant reflected rays are strongly elliptically polarised. At high angles of incidence, around 70–80°, and the correct angle of initial polarisation (near to but not quite 45°), the ellipticity becomes circular. Both these specific angles vary somewhat between different metals. The circular polarisation after linearly polarised light has been reflected from some shiny metal objects is shown in figure 8.10. It is also worth pointing out again that back-silvered glass mirrors also produce ellipticity but the effect is very small. Due to refraction in the glass, even light that is incident at a high, grazing angle has a much smaller angle of incidence on the metallic surface itself (figure 8.11). The maximum real angle of incidence with an ordinary glass mirror is about 41° which produces very little ellipticity.

It is now possible to see why the retardation colours seen in a mirror placed at 45° to the direction of polarisation, as described in chapter 2 and shown in colour plates 5 and 6, are different depending whether the mirror is front silvered or back silvered. The front-silvered mirror causes a degree of ellipticity in the reflected light that varies somewhat as the mirror is tilted to change the angles of incidence and reflection. The different ellipticity changes the interaction with the second polariser so that the reflected colours alter quite noticeably. Different metals also produce different combinations of polarisation and timing of retardation so that the colours may be strikingly different in two apparently similar cases. With an ordinary back-silvered mirror, the situation is simpler since little polarisation or ellipticity is produced: the angles of actual incidence and reflection (at the silver layer) vary rather little and cannot approach the high, grazing angles necessary to produce pronounced ellipticity.

The phenomena of dichroicity and birefringence, described earlier for their effects on linearly polarised light, also have their counterparts for circular (and therefore elliptical) polarisation. Circular birefringence, by analogy with its linear counterpart, is where left-handed and right-handed circularly polarised light are propagated at different speeds. This effect occurs in optically active materials: those with an asymmetrical structure or with chiral molecules in solution (see chapter 5). Indeed it provides an explanation for the rotation of the direction of linear polarisation seen in such cases.

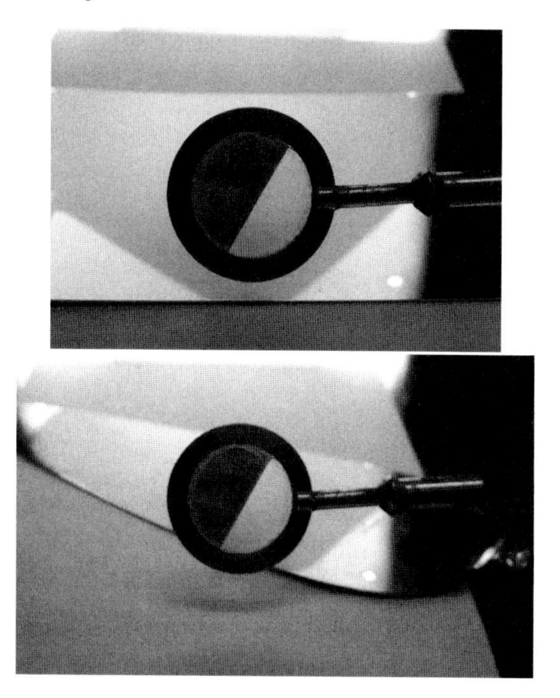

Figure 8.10. Reflection of linearly polarised light at high, almost 'grazing' angles by shiny metallic objects produces eliptical or even circular polarisation as shown by the Cotton polariscope of figure 8.7. The incident light is linearly polarised at about 45° to both the plane of incidence and the plane of the reflector (shadowy reflections of the polaroid can be seen). The reflected light is right circular or left circular, depending which of the two possible directions the linear polarisation occurs—i.e. it can be reversed by rotating the linear polariser by 90°. The two objects are a steel camper's mirror and a stainless steel cake slice.

Figure 8.12 shows that two rotating vectors of equal length and speed of rotation add up to a linear vector. This means that linearly polarised light can be regarded as equivalent to two components of circularly polarised light of equal strength and wavelength. Normally this concept is not very helpful and is best ignored but, as figure 8.13 shows, a change in the relative timing of the two rotating vectors when they recombine twists the resultant linear vector to either left or right. Since circular birefringence delays one circular component with respect to the other, it provides this change of timing and the consequent twist.

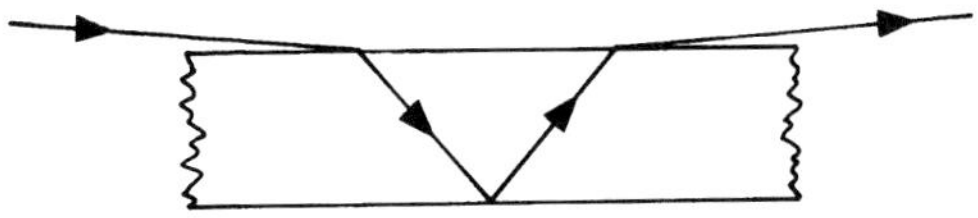

Figure 8.11. Normal back-silvered mirrors cannot reflect at high angles of incidence due to refraction by the glass. Light incident at a large, grazing angle is bent so that it is actually incident on and reflected from the shiny surface at about 41°. This is therefore the maximum angle of actual incidence and is insufficient to produce the circular polarisation seen in figure 8.10.

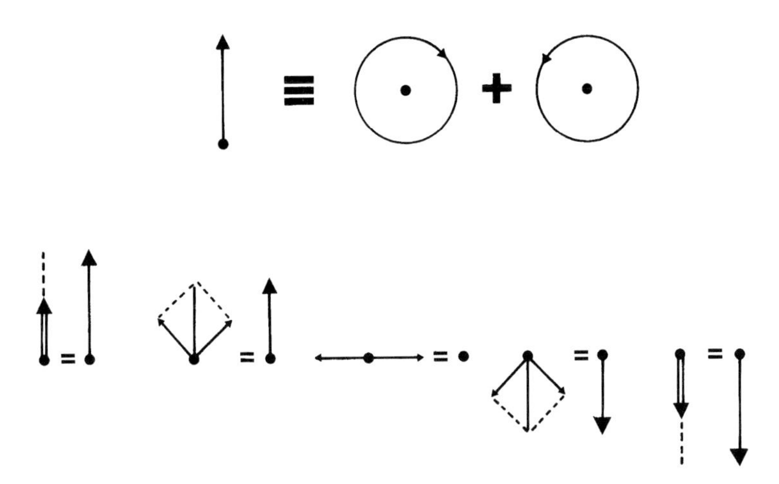

Figure 8.12. Two vectors of equal length, rotating in opposite directions at equal rates are exactly equivalent to one linear vector. This statement is summarised in the upper line and is demonstrated below by five instantaneous 'snapshot' figures at different times through half a rotation. As the two rotating vectors move apart, they add together to produce a single vertical vector that fluctuates in height and is normally summarised by the single arrow representing its maximum value in one direction only. The next half-rotation will restore the resultant to its original position and length. Dotted lines are for construction only.

This explanation was first put forward by Augustin Fresnel in 1822 and was proved by some ingenious measurements some 62 years later. A difference in refractive index means that rays will be bent to different degrees in a prism. The effect is quite small but a series of hollow prisms alternately containing left-handed and right-handed chiral solutions can

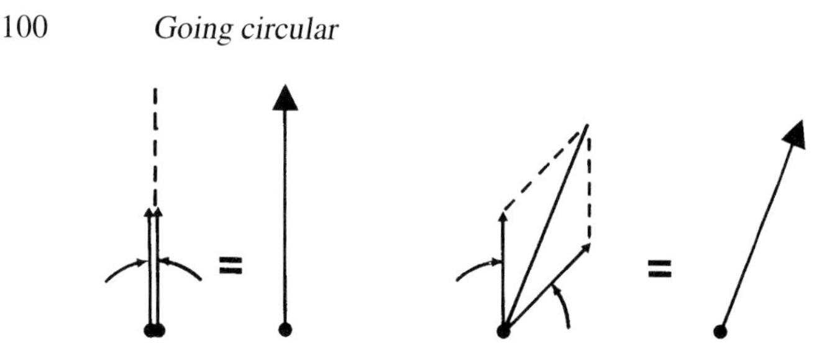

Figure 8.13. If one of the rotating vectors of figure 8.12 is delayed a little, they will meet up at a different point and so the resultant linear vector will be rotated. This shows that circular birefringence (when two circular vectors propagate at different speeds so that one of them is delayed) in sugar solutions rotates the direction of linearly polarised light as seen in the equivalent phenomenon of optical activity described in chapter 5.

actually be made to split linearly polarised light into two separate beams of circularly polarised light of opposite handedness.

Circular dichroism, sometimes called the Cotton effect, is where a material absorbs left-handed circularly polarised light more than right-handed, or *vice versa*, and transmits or reflects the remainder. It therefore forms a method of producing circularly polarised light from ordinary unpolarised light. Every fluctuating component of the light is resolved into two rotating vectors of which one is preferentially absorbed, leaving a net circularity in the opposite sense. This happens when the material contains asymmetrical molecules of one handedness. It also occurs in some liquid crystals (see chapter 3) and again depends on the presence of chiral molecules in which those of one handedness predominate, which often implicates a biological process somewhere in their origin.

An interesting natural case of circular dichroism is found in the wing cases and some other parts of certain brightly coloured or shiny chafer beetles, such as *Cetonia* (the rose chafer) and *Plusiotis*. These bright green insects absorb right-handed polarised light so that the colour they reflect is strongly left-handed. They look completely normal through a left circular polaroid but black through a right circular polaroid. This occurs because of a special structure in their external skeletons. These consist of birefringent layers of molecules that are all orientated in one direction within each layer. But successive layers are rotated steadily in a systematic helical arrangement, with the scale of the rotation being comparable with the wavelength of light (much larger than the size of the molecules). The effect of this arrangement has been called 'form dichroism' to distinguish it from 'molecular dichroism'. The phenomenon may be quite widespread as it is not visible to the eye and can only be detected by use of a circular ('Cotton') polariscope such as shown in figure 8.7. It emphasises the point that there is still much scope for scientific discovery based on simple observation, given some diligence and a little ingenuity.

Chapter 9

Seeing the polarisation

From all that has been discussed in the earlier chapters, it seems that nearly all the natural daylight we see is at least partially polarised. Polarisation is unavoidable because we never look directly at the sun and all the light that actually enters our eyes has been either reflected or scattered by something. It turns out that many animals are able to detect both the degree of linear polarisation and its direction, and they exploit this information in several ways. Such creatures include insects, crustacea, octopus and cuttlefish and some vertebrates but not, except in an insignificant way, ourselves and other mammals. Having already seen in this book what we are missing, it is pertinent to consider how other eyes respond to polarisation and why ours do not.

It has long been suspected that some animals are sensitive to the direction of polarisation because that would explain some of the things they do. But early tests proved negative until some experiments were published by Irene Verkhovskaya in Moscow in 1940. She studied *Daphnia*, the small freshwater crustaceans popularly called 'water fleas', which migrate vertically over several metres in the water each day. She found that they are attracted to light and gather together where it is strongest. In several experimental arrangements, linearly polarised light was found to be two to three times more effective than unpolarised light in attracting the animals. At dawn, when they move towards the surface, the sky overhead is maximally polarised unless it is overcast, so this might increase the stimulus to migrate upwards. It was more than a decade later, however, before others showed that *Daphnia* are able to detect the actual direction of polarisation, since they align their bodies, and so direct their paths, at right angles to it.

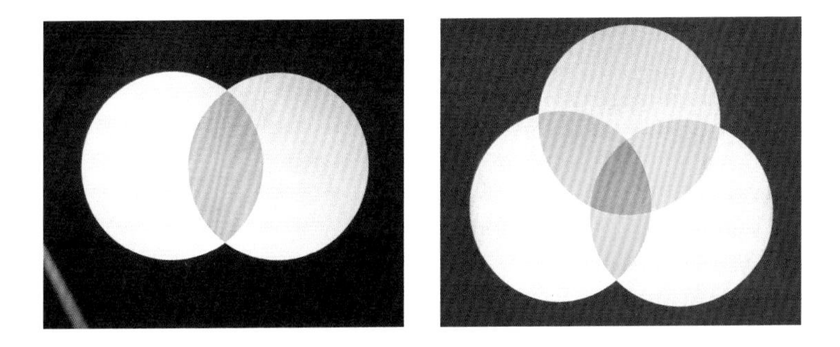

Colour plate 1. Two jam-pot covers overlaid in the same orientation give double the retardation—in this case from 235 to 470 nm, giving an orange colour between crossed polarisers. Three such films give 705 nm and a blue colour. By matching the colour with a known scale (e.g. colour plates 4 and 18) the retardation can be easily and accurately measured and that of each single film can then be deduced. (See pp 10–11.)

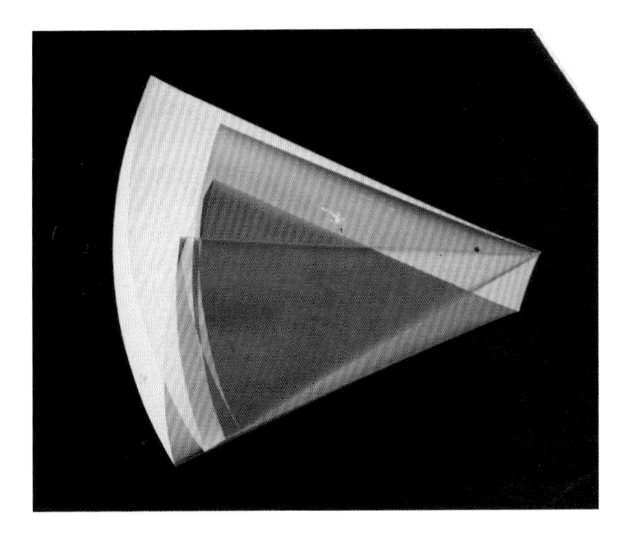

Colour plate 2. A jam-pot cover that has been folded at random and viewed between crossed polarisers. The different amounts of overlap and changed orientations produce a variety of retardations and therefore of colours as seen between crossed polarisers. (See p 11.)

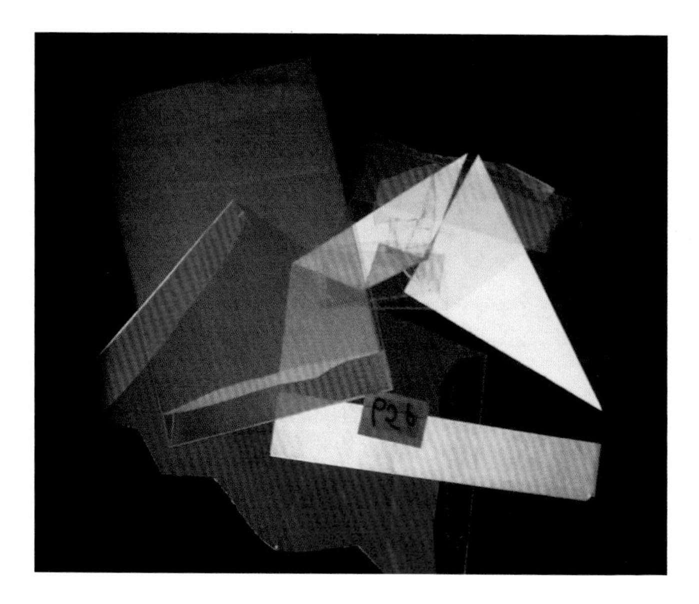

Colour plate 3. Different thicknesses of cellophane film taken from greetings card wrappers and chocolate boxes provide a variety of retardations and so produce strikingly different colours when scattered at random between crossed polarisers. (See p 12.)

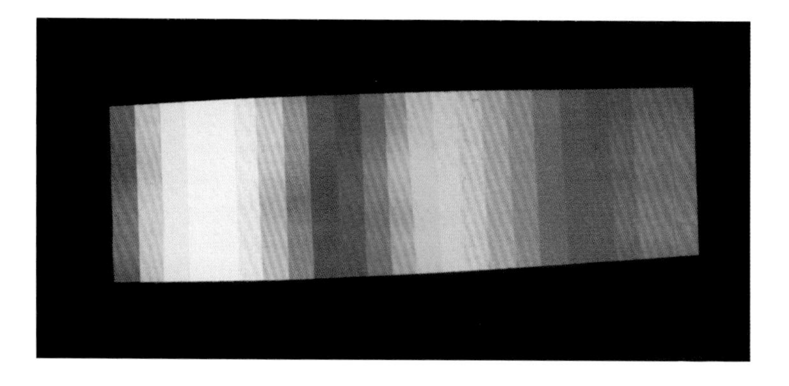

Colour plate 4. A birefringent 'step-wedge' which gives increases in optical retardation in 22 steps of 55 nm from 0–1210 nm. It was made by applying successive layers of clear plastic tape. Between crossed polars it shows just over two orders of Newton's colours, or retardation colours. (See p 13.)

Colour plate 5. Three discs of retarder films have been mounted in supporting rings and are here viewed between crossed polarisers. A mirror at 45° to the direction of polarisation shows reflections of the discs that are of different colours. If a polished metal mirror is used instead of the silvered glass used here, the colours may be different again. (See pp 14, 97.)

Colour plate 6. The optical retardation step wedge from colour plate 4 together with its image in a mirror held at 45° to the direction of crossed polarisers, giving a more comprehensive illustration of the effect seen in colour plate 5. (See p 15.)

Colour plate 7. A polariscope that produces colour contrasts can be made by mounting strips of polaroid of alternating orientation and backing them by a cellophane film giving about 650 nm retardation; the blue and yellow colours seen in polarised light are interchanged when the device is rotated by 90°; other colour pairs may be preferred and a range of retarder films can be tried. If the same device is reversed so that the film is toward the viewer instead of towards the light source, it produces a black–white contrast instead of colour. (See p 16.)

Colour plate 8. Kaleidoscope images produced from clear cellophane scraps of various thicknesses loosely held between two sheets of polaroid. Rotating one of the polaroids changes the colours and rotating a cellophane film that crosses the whole field modulates these colours to offer a new range of possibilities. (See p 16.)

Colour plate 9. Some examples of photoelastic strains 'frozen' into clear polymer resin during the manufacture of common objects and made visible between crossed polarisers. (See p 17.)

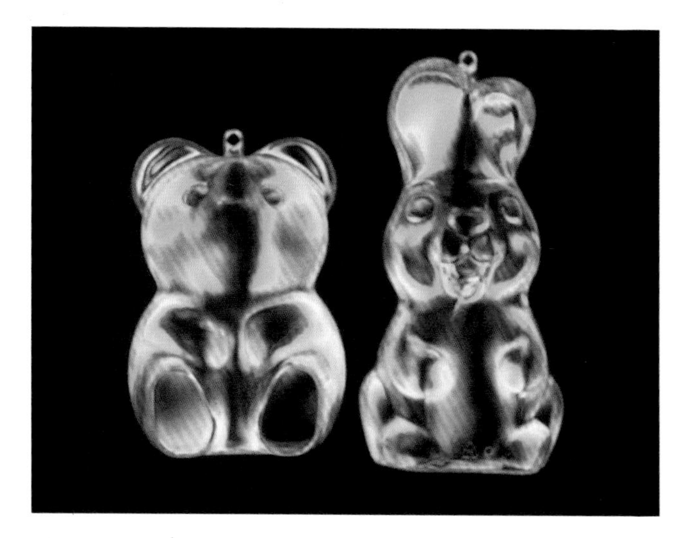

Colour plate 10. A further example of simple clear plastic objects that are transformed when viewed between crossed polarisers: containers for Easter chocolates around 10 cm in size. (See p 17.)

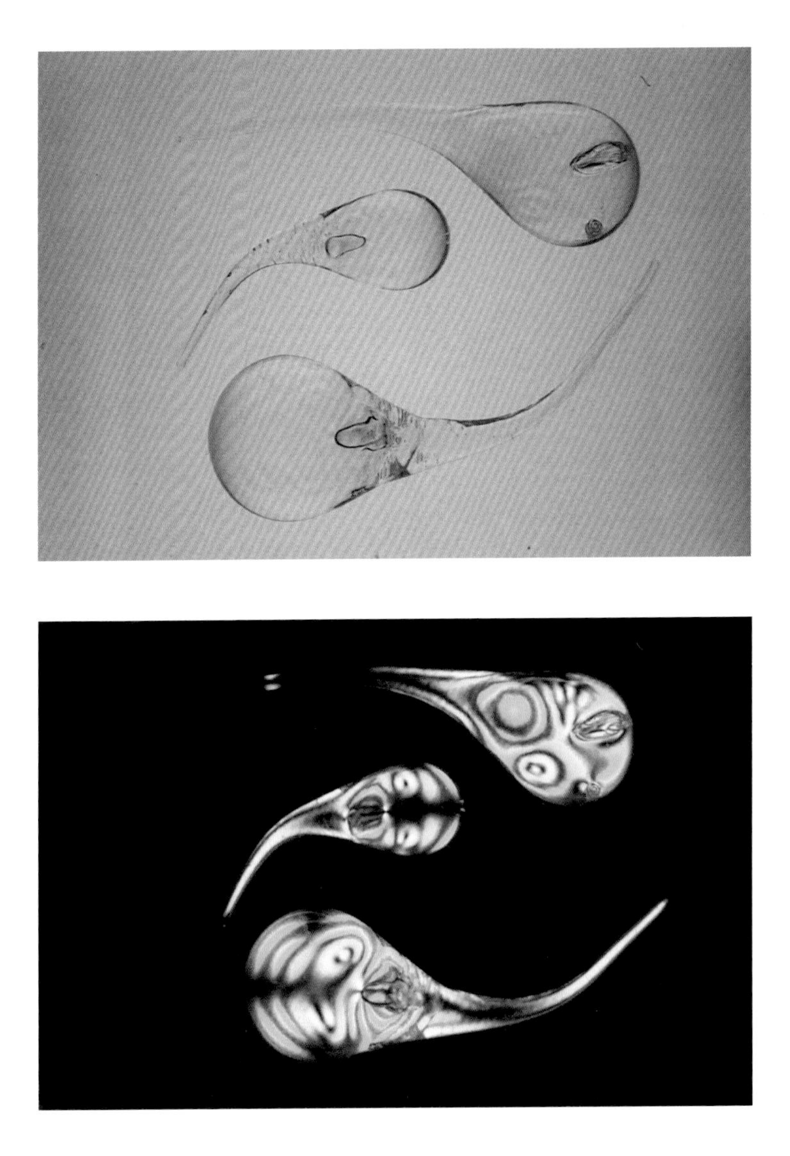

Colour plate 11. Prince Rupert's drops made by dropping molten glass into cold water. They are 1–1.6 cm in diameter. When viewed between crossed polarisers they show many closely packed coloured fringes due to the very high internal strains. (Kindly made by John Cowley, Glass Workshop, Queen Mary, University of London. (See p 19.)

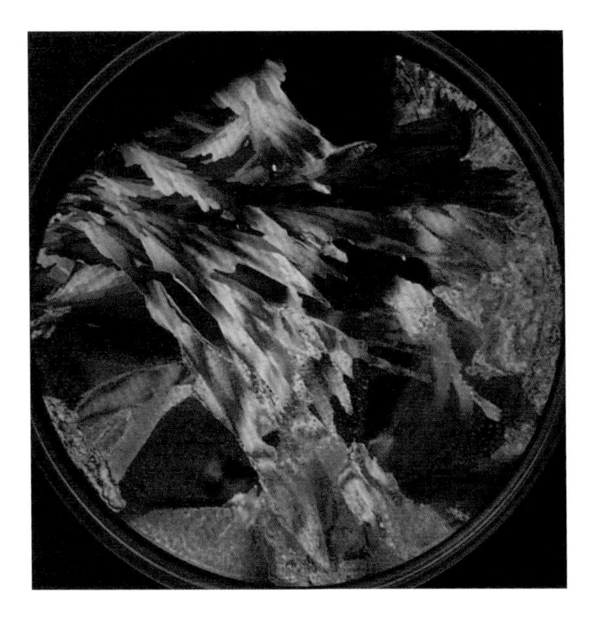

Colour plate 12. A shallow dish of water ice as seen between crossed polaroids. Each of the individual crystals in the mass produces a colour that depends on its orientation and thickness. (See p 21.)

Colour plate 13. Thin salol crystals, formed between two sheets of glass, as seen between crossed polaroids. The directions of growth of the crystals shows vividly. (See p 21.)

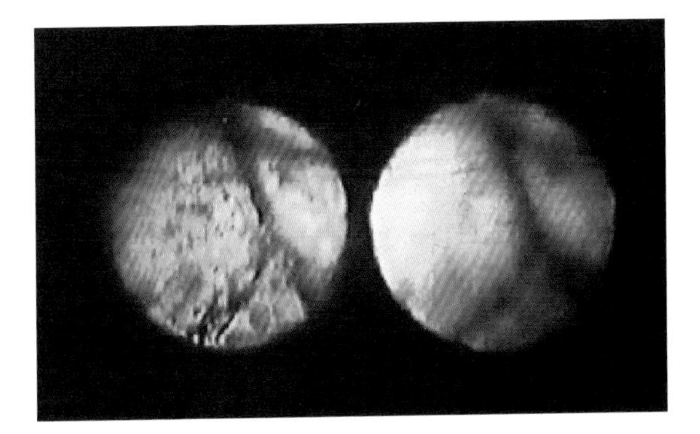

Colour plate 14. A dichroic copper acetate crystal as seen through the dichroscope of figure 3.6. Coloured glass or any other non-dichroic material would always give two images that are identical to each other, except for their polarisation. Because of the difference in refractive index, the two images appear at slightly different distances through the crystal, so one is slightly out of focus. (See pp 23, 25.)

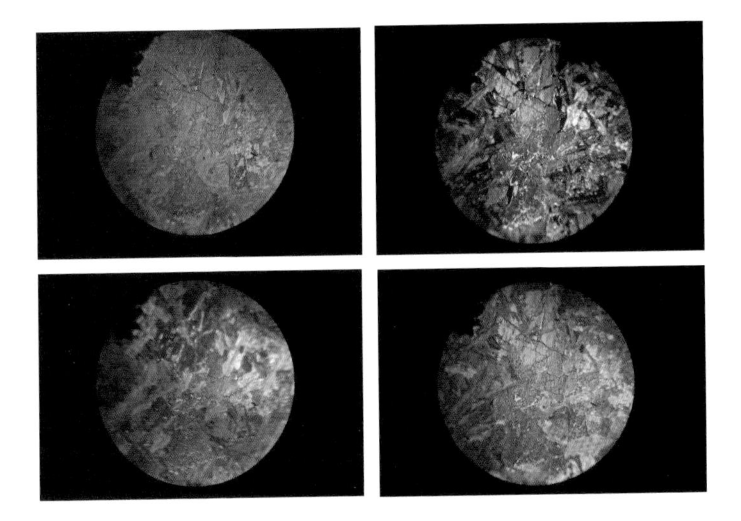

Colour plate 15. A section of igneous rock ground to a thickness of about 30 μm (thousandths of a millimetre) and viewed through a low power microscope: upper left, the appearance in normal transmitted light; upper right, the same field between crossed polaroids; below, the same again but with superimposed retarder films of about 450 nm (left) and 750 nm (right). (See p 34.)

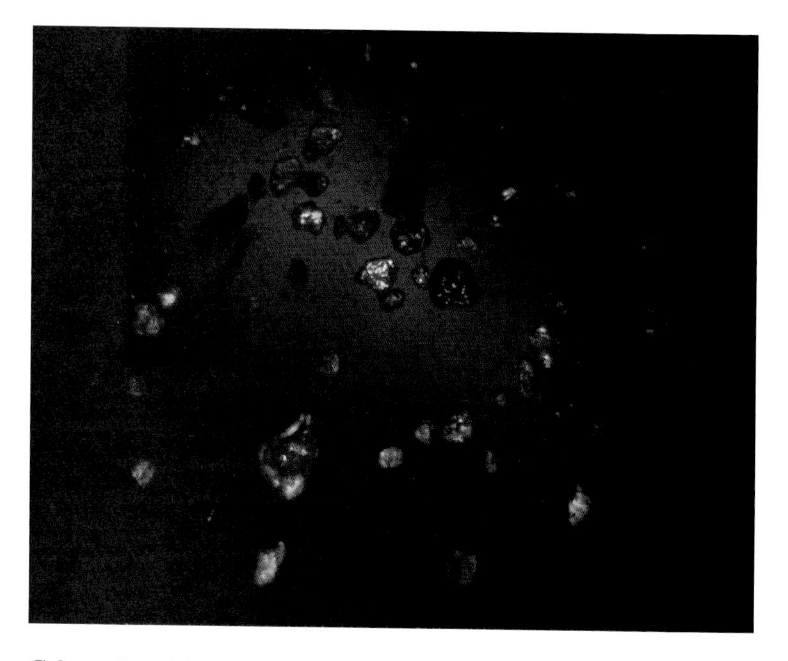

Colour plate 16. Grains of silver sand as seen between crossed polarisers through a low power microscope. The varying colours are due to irregular paths for light passing through the crystals. Compare these with the much larger quartz crystals of figure 3.2 which appear colourless between crossed polarisers. (See p 34.)

Colour plate 17. Thin sheets of mica showing retardation colours between crossed polaroids and the complementary colours between parallel polaroids. These sheets were extracted from a rather large electrical capacitor kindly donated by Ken Edwards. (See p 35.)

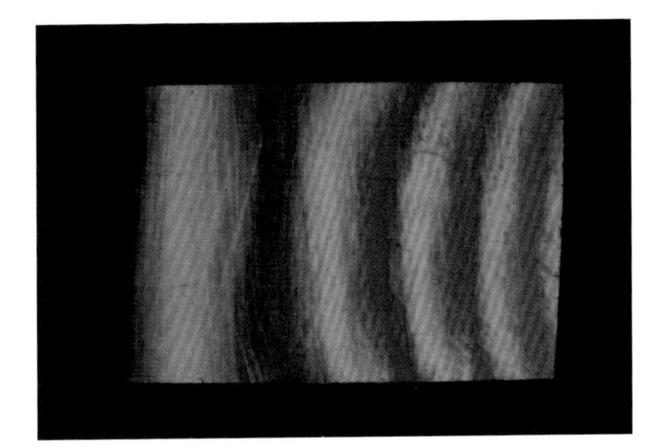

Colour plate 18. An improvised optical retardation wedge made by gently grinding a gypsum flake on an abrasive stone, and finishing off with silver polish. The crystal was initially about half a millimetre thick, having been cleaved with a sharp knife from a larger piece. After mounting, it was covered by a half-wave film which was turned so as to subtract its retardation from that of the crystal in order to get low values of retardation at the thin end (black is zero). The finished device is here shown between crossed polarisers, with several orders of retardation colours. (See p 35.)

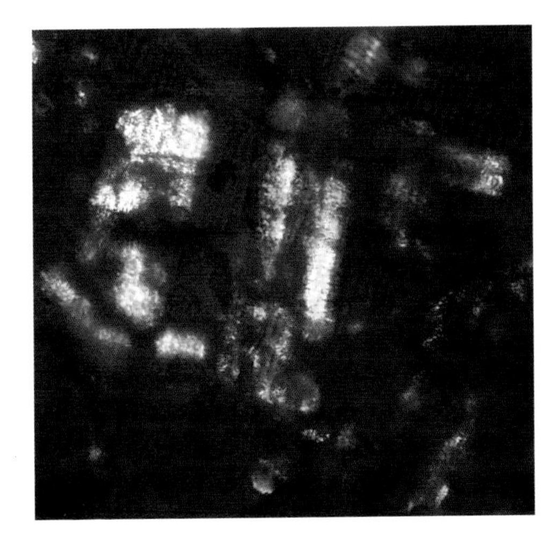

Colour plate 19. Hair has a birefringent crystalline structure and shows retardation colours between crossed polarisers. Here is some chaff from an electric razor, seen under a microscope. (See p 36.)

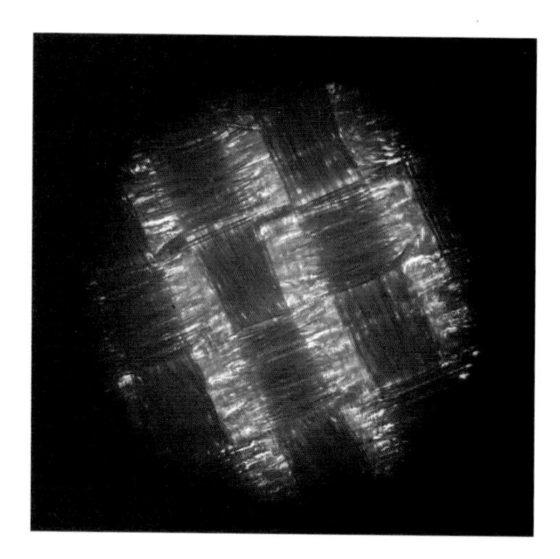

Colour plate 20. Silk has a birefringent crystalline structure. Here a piece of a white silk handkerchief shows retardation colours when photographed between crossed polaroids under a microscope. (See p 36.)

Colour plate 21. Cotton is a vegetable fibre that shows birefringence and thus retardation colours between crossed polarisers. Here the end of a white cotton thread is frayed out and photographed under a microscope. (See p 36.)

Colour plate 22. The scattering produced when a little Dettol is stirred into water that is lit from below. The turbidity produced is rather blue in colour because the very tiny droplets of Dettol scatter short wavelength blue light more strongly than longer wavelength red light. Instead of Dettol a little milk can be used or fine sulphur particles can be produced by adding dilute acid to hypo solution (sodium thiosulphate). With sulphur the scattering becomes less blue as the particles grow larger and begin to reflect all colours more equally. In the same way a fine mist appears blue but the larger water droplets in clouds make them look white. If there is much blue scattering, the light that emerges from the top of the column, seen reflected on a white card (as in figure 6.2), appears strongly reddened because the shorter wavelengths have been selectively eliminated. This simulates a red sunset. (See pp 61, 64.)

Colour plate 23. When a polariser and a retarder film are introduced below the scattering column, retardation colours for crossed polarisers are produced on opposite sides of the column with the complementary (parallel polariser) colours at right angles in between. Here both views are shown for each of two retarders: about 370 nm (top) and about 550 nm (bottom). The strong blue scattering appears even brighter when other colours are removed as it is now more saturated. The purple and orange colours are less bright but they do show that some of the longer wavelengths of the light are scattered too. (See p 62.)

Colour plate 24. When strong sugar solution is used in a scattering column above a polariser, the scattering is twisted along the column. But different wavelengths are twisted at different rates and soon become noticeably separated, giving a 'spectral barber's pole' effect. Rotating the polariser rotates the scattering and the colours appear to screw their way up the pole as others come into view. This shows again that, whereas small particles preferentially scatter the shorter blue wavelengths, they also scatter longer wavelengths to some extent, just as the blue sky actually has all colours of the spectrum in its light. (See p 62.)

Colour plate 25. Primary and secondary bows in the spray from Victoria Falls. Both bows lie at the edges of white scattering areas while the band between them is clear so that the background is not obscured. (See p 82.)

Colour plate 26. The Bachalpsee in Switzerland at dawn. The distant sharp peak is Finsteraarhorn on a bearing of 155° and the nearer peak to the left is Schreckhorn. The clear blue sky, whose light must be strongly vertically polarised, is not reflected in the nearer part of the lake where the water can only reflect horizontally polarised light. (See p 85.) (By courtesy of Ernst Schudel, Photo-Suisse, Grindelwald.)

Colour plate 27. The Taj Mahal and reflecting pool looking due north at sunset. The nearer water does not reflect the blue sky although it reflects the building very clearly. (See p 86.) (By courtesy of Abercrombie and Kent Travel.)

Very recently the preference of *Daphnia* for polarised light has been confirmed and has been linked to the tendency of these and other small invertebrates to avoid the vicinity of the shore. Light penetrating into deeper water is scattered (see chapter 6) and, so becomes horizontally polarised, much more than light falling onto shallows near the beach. So the animals are attracted away from the dangerous shallow regions until they are surrounded by safer deeper water. The two basic compound eyes of *Daphnia* are fused together to form a single median eye, with a total of 22 facets, that swivels rapidly around its axis within the head. The rotating eye presumably scans all possible directions of polarisation, perhaps to ensure that it detects the predominantly horizontal polarisation of its environment, as the animal swims along in varying attitudes. This would account for the apparently equal attractiveness of different directions of polarisation first seen in the original experiments.

The early experiments on *Daphnia* showed they are sensitive to the existence of polarisation but they were not the first animals shown to be able to detect the actual direction of polarisation. This was first demonstrated in bees by the great pioneer of animal behaviour studies and Nobel Laureate Karl von Frisch in 1948. He had already found that worker bees can communicate to other bees the direction and distance of a source of food, either nectar or pollen, by an excited dance performed repeatedly on the surface of the comb in the hive (figure 9.1). The direction is indicated by the 'waggle run' element of the dance in which the bee runs forward waggling her abdomen before turning alternately left and right to return to the start. The inside of most hives is completely dark, the combs are vertical and 'upwards' is regarded by the bees as representing the direction of the sun at the time; the waggle run then deviates left or right from the vertical according to the bearing of the food source to left or right of the sun itself (figure 9.1). As the sun moves across the sky through the day, so the waggle run for a particular direction gradually rotates appropriately, at 15° per hour or half the rate of the hour hand on a clock face.

When von Frisch experimentally abolished the vertical reference by arranging the combs horizontally, the waggle runs then pointed in the actual direction of the food source itself, but only if the bees could see the sun or part of the sky through a window in the roof of their hive. The bees could just cope when only 10–15% of the sky was visible to them, provided it was blue; when their field of view was covered by a cloud, or when their window was covered over, the dances immediately became

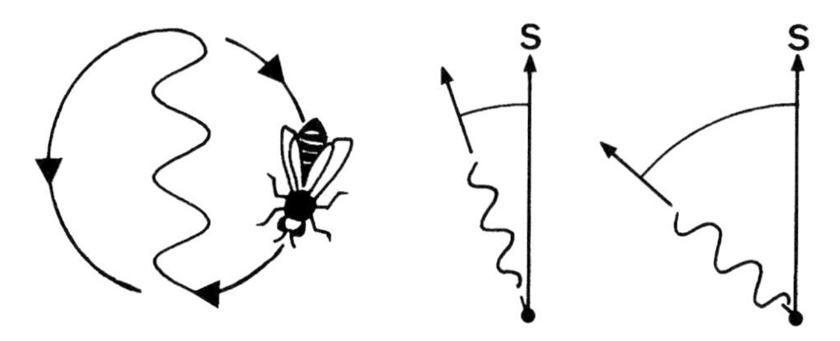

Figure 9.1. The waggle dance of the honey bee. Left: the pattern of the dance, turning alternately left and right with the waggle run up the middle. The angle of the waggle run from the vertical indicates the bearing of the food source with respect to the sun and the number of waggles gives the distance to the food. Right: the waggle run for a source 20° to the east (left) of the sun and the same two hours later when the sun has moved 30° westwards.

disorientated and random. The fact that a portion of blue sky provided a sufficient reference when the sun itself could not be seen was a major surprise but von Frisch suspected that the bees might be able to detect the sky polarisation pattern that is itself determined by the sun, as described in chapter 6. To test this he covered the window in the hive with a large piece of polaroid. When he turned the polaroid so that the polarisation of a patch of sky was altered in a controlled way, the bees' dances rotated accordingly. The story has been vividly told in several popular books by von Frisch himself, translated into English, and it is well worth reading in greater detail.

In order to understand how the bees distinguish the direction of polarisation, it is necessary to know something about how their eyes work. The principal paired eyes of insects are of the type known as compound eyes and have a quite different action from the familiar simple or 'camera' eyes of vertebrates such as ourselves. In the latter, as in a camera, a single lens (generally combined with refraction by the curved cornea) focuses an image on the retina, which is a screen of light sensitive cells. Nerve fibres connect each point of the retina with the brain and thus transmit a detailed representation of the image. By contrast a compound eye consists of a convex array of separate facets or optically independent units, about 5500 in each eye of a worker honey bee, all pointing in different directions. Each unit, called an ommatidium, has its own lens which is hexagonal in outline so that they pack together to collect all the available light (figure 9.2). Behind

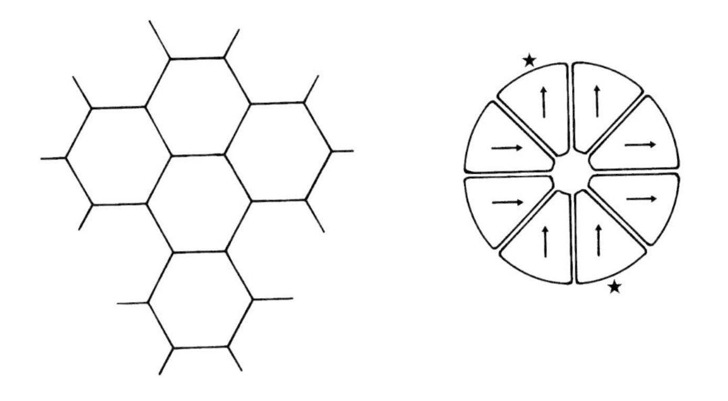

Figure 9.2. The compound eye of the honey bee. Left: the outer surface showing part of the array of hexagonal lenses, each of which faces outwards in a slightly different direction. Right: below one lens (at greater enlargement) showing the arrangement of the eight main light sensitive cells in cross section. The arrow within each cell shows the polarisation direction to which it is most sensitive. The two cells marked by stars respond best to ultraviolet light and the rest to either blue or green light.

each lens there are essentially eight light-sensitive cells (figure 9.2) in an octagonal pattern, each connected to the brain by a nerve fibre. This little set of cells does not qualify for the term retina and is called a retinula or 'little retina'. Clearly with such limited attributes one ommatidium cannot form and analyse an effective image and the visual field is only represented by the combined action of the whole array of such units.

Now von Frisch imagined that each of the eight light sensitive retinula cells might somehow be sensitive to one direction of polarisation so that one ommatidium would be able to analyse polarisation in terms of four components 45° apart (figure 9.3). This idea was correct in principle although the details turned out to be slightly different. Actually four of the eight cells are potentially most sensitive to one direction of polarisation while the other four are most sensitive to light polarised at right angles to this, giving simultaneous analysis into two orthogonal components (figure 9.2). The position is complicated, however, by the fact that the same cells also serve for colour vision which, as in our eyes is achieved by comparing the excitation of three kinds of cells with different spectral responses: in the bee retinula four of the eight cells are most sensitive to green, two to blue and two to ultraviolet (figure 9.2).

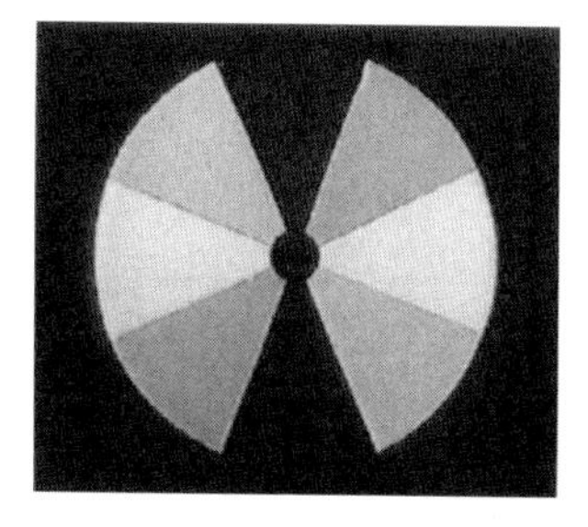

Figure 9.3. Karl von Frisch imagined the main eight sensitive cells of a bee's ommatidium might be sensitive to four directions of polarisation. He made a model from pieces of polaroid which makes a very good polariscope as shown in this replica. The actual polarisation sensitivities as determined by later, physiological studies, are shown in figure 9.2.

In common with most other insects, bees have no red-sensitive cells. It is the ultraviolet cells that are most responsive to the direction of polarisation although an additional short ninth cell, also an ultraviolet unit and tucked in among the bases of the other eight, may be the most polarisation sensitive of all in most ommatidia.

The basis of this polarisation sensitivity is that in these eyes the light absorbing process is inherently dichroic. The visual cells of all animals detect light when it is absorbed by special receptor molecules that are thereby changed in shape by a photochemical reaction; this initiates a chain of events that culminates in messages to the brain via nerve fibres. The sensitive molecules, of substances called visual pigments, are elongated and absorb light most readily when the direction of polarisation is aligned with their long axis, or at least with a series of carbon=carbon double bonds within the molecules. In the light sensitive retinula cell of an insect compound eye these pigment molecules are all orientated in the same direction. They are held in the membranous walls of minute parallel tubular pockets called microvilli, that are tightly packed together and run from the wall of the cell across the light path (figure 9.4). Thus all the absorbing molecules lie across the light path and, with their axes all more or less parallel, the system as a whole is dichroic, like some crystals. Light polarised in one direction is strongly absorbed, leading to excitation of the cell, while light polarised at right angles is not absorbed and has no effect.

One final complication in the bee's eye concerns twisting of the

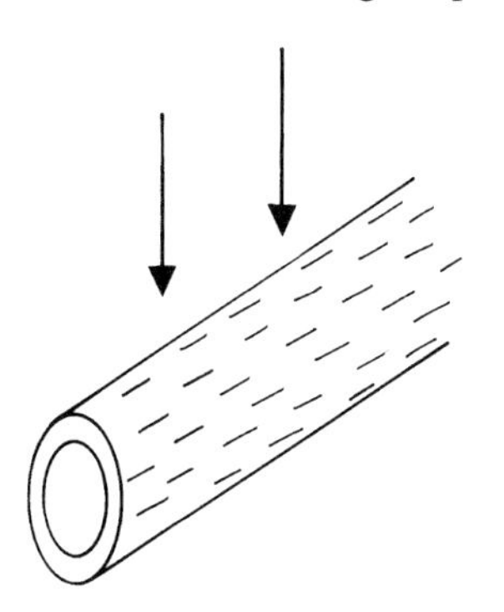

Figure 9.4. The sensitive cells of the eyes of a bee and many other invertebrate animals contain tiny tubes called microvilli that lie across the light path (arrows). The membranous walls of the microvilli contain the light sensitive pigment molecules (represented as dashes) which are all orientated in the same direction. This almost crystal-like regularity makes the light absorbing process dichroic so that the cell is inherently most sensitive to light polarised along the axes of the microvilli.

group of retinula cells. In the system described earlier, there could be an ambiguity between colour and polarisation: both are detected by the brain comparing the excitation of different cells within the ommatidium, but differences of response may equally be due to the colour of the light or to its polarisation. In order to avoid this problem, the bundles of sensitive cells in most ommatidia are twisted along their length by more than 180°, like strands of a rope, clockwise in some ommatidia and anticlockwise in others. Thus the axis of dichroicity rotates so that light not absorbed in the outer part of a cell may well be absorbed when it passes further down. This improves sensitivity because the total light capture is increased, but although the membranes with their visual pigment molecules are actually dichroic, the twisted cell as a whole is not, or it has greatly reduced susceptibility to polarisation. This may explain why the short ninth cell is often the most sensitive to polarisation, because its very shortness means it is never twisted enough to destroy its inherent dichroicity. The non-dichroic cells, however, are ideal for unambiguous colour analysis because differences in their excitation can only be due to differences in their spectral responses.

However, there is a band of about 150 specialised ommatidia that lie in the dorsal region of the eye and so face upwards. They are called the

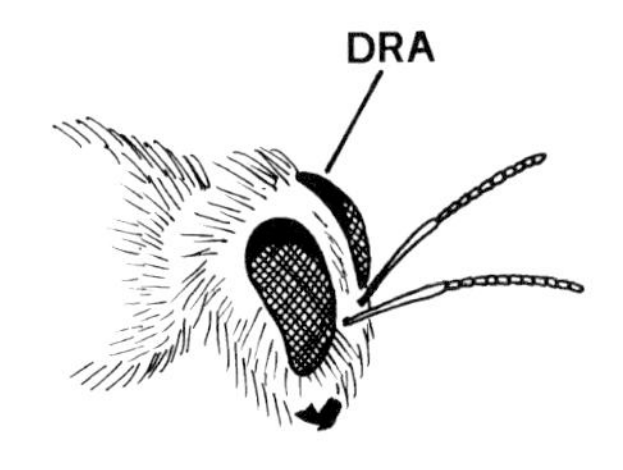

Figure 9.5. The eye of the honey bee showing the dorsal rim area (DRA in black) where about 150 upward-facing ommatidia contain polarisation-sensitive ultraviolet-sensitive cells, responsible for analysing the polarisation pattern of the sky.

dorsal rim units (figure 9.5) and they have no twist to their retinula cells. The cells are therefore highly dichroic as a whole and are thus ideally adapted to perform rapid assessment of the polarisation pattern of the sky (see chapter 6). It is this band of dorsal rim ommatidia that is responsible for the sky compass navigational abilities of the bee. Because they point upwards they are unlikely ever to be used for discriminating colours, of flowers for example, and the possible ambiguity described previously is therefore irrelevant. In some of these ommatidia the ninth cell is not short but runs up the bundle with the others, while the microvilli, with their dichroic pigment molecules, may be arranged not in just two orthogonal directions in different cells but in three or even more directions within the retinula of one ommatidium.

A different and rather more 'extravagant' solution to this problem of colour/polarisation ambiguity is found in the remarkable eyes of mantis shrimps, *Squilla*. Most crustacea, like insects, have compound eyes. Those of mantis shrimps contain as many as ten different sensitive pigments so their colour vision must be quite outstanding. Perhaps because of this variety, in the mid-band of the eye there are horizontal rows of ommatidia that 'look at' the same regions of the visual field, instead of all pointing in different directions. Two rows of these mid-band ommatidia are sensitive to polarisation in the broad range from blue to yellow light. So in these animals colour vision and polarisation analysis are served by different specialised ommatidia that face in the same direction. The function of polarisation sensitivity in mantis shrimps is as yet unknown.

Ants also show discrimination of sky polarisation. Some especially revealing studies have been made on desert ants, *Cataglyphis*, of North

Africa and the Middle East. These insects must forage across open spaces, often with no visible landmarks to guide them. Even if their outward path is quite tortuous they can still take a direct line back to their nest over as much as 200 m. Furthermore, unlike bees, they can be followed by experimenters who can manipulate their visual environment all the way by devices held over the walking ants. For instance the sun can be obscured by a shadowing card, leaving most of the sky clearly visible, or conversely the sky polarisation can be changed by a large sheet of polaroid while leaving the sun visible through it. It has been known since 1911 that some ants, such as the garden ant *Lasius*, rely predominantly on the sun—using so-called sun compass orientation. Shielding them from the sun, and at the same time showing them a reflection of the sun in a mirror, makes them change or even reverse their direction of travel accordingly. But the desert ant relies more on the pattern of sky polarisation than on the sun itself. In this case each eye has about a thousand ommatidia of which about 80 dorsal rim units are specialised polarisation-sensitive, ultraviolet units. Reliable navigation back to the nest without delay is essential in the desert where long exposure to the hot sand can be lethal. Perhaps a small cloud over the sun could impose a dangerous delay if only the sun compass were used.

Polarisation sensitivity of this kind now seems to be very common among the insects and examples of a different kind of application by dragonflies and water boatman bugs have been given in chapter 7. As in the bees and ants, it is generally especially associated with the ultraviolet-sensitive cells. An exception is seen in crickets where the dorsal rim cells are all blue sensitive dichroic units although ultraviolet- and green-sensitive cells occur elsewhere in the eye. The probable explanation is that whereas bees and ants are active by day, crickets tend to be nocturnal, when levels of ultraviolet light are low. It has been suggested that this may be common in other nocturnal insects although some diurnal flies also seem to be most sensitive to polarisation at blue wavelengths. Strangely, a few insects, such as certain water beetles, show best polarisation analysis at the (to them) very long wavelengths of yellow–green light and this has not yet been very convincingly explained. In general, the main significance of polarisation analysis seems to be the ability it confers of using the sky pattern for orientation, not just for homing as in bees and ants but also for keeping to a straight track. Many different insects, including butterflies, flies and mosquitoes, appear to be reluctant to fly across open spaces when the sky is heavily overcast. Even on a sunny day, it is not possible to keep the sun in sight when moving through a wood whereas patches of sky can nearly always be seen here and there through the canopy.

The presence of microvilli with aligned receptor molecules, and their orthogonal orientation in different cells of the ommatidium is common to insects, crustacea (including *Daphnia*) and many other arthropods in general. There are some variations, for instance bunches of microvilli of different cells may actually interdigitate in an orthogonal 'dovetailed' arrangement so that light passes through each in turn along the axis of the cells. But the basic arrangement is ubiquitous. In a number of cases it has been possible to observe dichroicity within individual receptor cells by measuring light absorption for different directions of polarisation. It is tempting, therefore, to suppose that all these animals are able to detect the direction of polarisation. But this may not always be so. Twisting of the bundle of receptor cells has already been mentioned as a reason why the basic dichroicity may be lost—and twisting is not easy to observe as it requires cell orientation to be followed through a lengthy series of electron microscope sections. In many compound eyes the retinula cells within an ommatidium may also be coupled together electrically so that their excitation is shared. This increases the overall sensitivity to light but it destroys any individual dichroicity. Any such coupling through the membranes of adjacent cells cannot be seen under the electron microscope and is only revealed by electrophysiological recording of the responses of cells to light.

Actual evidence for polarisation analysis can take a number of forms, either behavioural or physiological. Examples of spontaneous behaviour in response to polarised light, and of changes of behaviour when the direction of polarisation is manipulated experimentally, have been described already. A number of investigators have also trained animals to respond to the direction of polarisation in an artificial visual stimulus. For instance an animal may be rewarded for pressing a pedal when shown vertically polarised light but not when shown horizontally polarised light. Such experiments, however, may be misleading even when the responses appear to be completely reliable. The problem is that an animal working for a reward will identify any clue as to which is the 'correct' signal and it may not be the clue the experimenter has in mind. This is especially true of polarised light stimuli. For instance, the light is sure to be reflected from some part of the enclosure or from objects within it. But light polarised in one direction is often (indeed generally, to some extent) reflected more strongly than light polarised at right angles (chapter 7). So an animal that is completely insensitive to polarisation might nevertheless spot differences in the brightness of a reflection and learn to respond quite consistently to achieve a highly 'correct' score.

A more direct way of demonstrating sensitivity to the direction of polarisation is to record the excitation of individual receptor cells or their messages to the brain in single nerve fibres. A very fine electrode is inserted into a receptor cell or nerve fibre and is used to record the electrical changes brought about by light stimuli. When polarised light is used, the direction of polarisation is varied to give the strongest response, and this is compared with the response to light of equal intensity polarised at right angles to the first. A large difference means that the response is greatly dependent on polarisation and therefore the animal as a whole could be affected; a small difference of response naturally means the unit is only weakly sensitive to the direction of polarisation. Then if two responses with orthogonal directions can be found in different units, the animal should be able to analyse the direction of polarisation. This approach has now been used successfully to study bees, ants, crustaceans and a range of other creatures .

Incidentally, it seems to be commonly assumed that two orthogonal sensitivities are necessary to analyse the direction of polarisation but of course one strongly sensitive direction will do provided that the animal is able to rotate its head, or at least its eye, and make successive observations. Conversely, even two orthogonal sensitivities can give ambiguous answers if no movement is made—vertically and horizontally sensitive cells will be equally stimulated by oblique polarisation at 45°. The problems are exactly the same as with the polariscopes described in chapters 1 and 2. In the case of reflections from water, which is always horizontal, no such ambiguity need arise and two orthogonal sensitivities will do for the water boatman bug or the dragonfly (chapter 7). But the sky compass analyser of the dorsal rim of the bee's eye has a more difficult task that justifies a greater complexity, with more than two sensitive directions.

Octopus, squid and cuttlefish are very highly developed molluscs. Their eyes are not multifaceted compound eyes but simple or 'camera' eyes, optically very like our own although they have evolved quite independently. The receptor cells of the retina contain their receptor molecules within microvilli and half the cells have their microvilli running horizontally while the other half have them vertically. Two cells of each kind form tetrads throughout the retina, suggesting the presence of polarisation sensitivity. Recently, behavioural tests have shown that an octopus can indeed analyse the direction of polarisation. They have been trained to respond to lamps that have a polarisation contrast pattern, say a vertical polarisation in the centre with a horizontal polarisation in the surround, or *vice versa*. They can even respond when the difference of direction is as small as 20° instead of 90°. The risk of 'false' cues being involved here can be made very small. Moreover, octopuses are also able to distinguish between a clear piece of Pyrex glass with no strains and an otherwise identical piece that has been subjected to heat stress to create internal strains. These produce birefringence that is only made visible by polarisation analysis (see chapter 2 and figure 2.10)—both glasses look identical to our eyes.

The significance of all this for vision in their environment is not yet clear but perhaps it enhances their ability to detect the shiny fishes on which they prey. Silvery scales reflect the colour of their surroundings, which is perfect camouflage underwater, but the reflections can be strongly polarised in ways that do not match the polarisation produced by scattering in the surrounding water. So polarisation analysis can be used to 'break' the camouflage. Quite apart from this, some cuttlefish apparently use polarisation to communicate with each other. As well as signalling by their well-known transient patterns of light, shade and colour, they can also produce patterns of light polarisation in their skin by means of iridophores: cells that are iridescent because they produce multiple internal reflections which interfere and polarise the light they reflect (see chapter 7). Like all iridophores, these cells are not themselves changeable but in cuttlefish they can be quickly concealed or exposed by tiny overlying sacs of black pigment controlled by radial muscle fibres. It has been suggested that these cuttlefish are thus able to communicate 'secretly' with each other without disturbing their general camouflage patterns of shade and colour that alone would be seen by any predator that lacks polarisation sensitivity.

Spiders are much more closely related to insects and crustacea, since all three belong to the Arthropoda, or joint-limbed animals with external skeletons. Yet their eyes are of the simple or 'camera' type (actually most insects also have three simple eyes called ocelli, but they are quite crude). The eyes of spiders vary considerably but most have four pairs of eyes looking in somewhat different directions. In the hunting or 'wolf' spiders, *Arctosa* (figure 9.6), the two 'principal', or anterior median eyes have large lenses and form good images. They are used for orientation and respond to the polarisation pattern of the 'sky compass'. Both they and the upward-looking posterior median

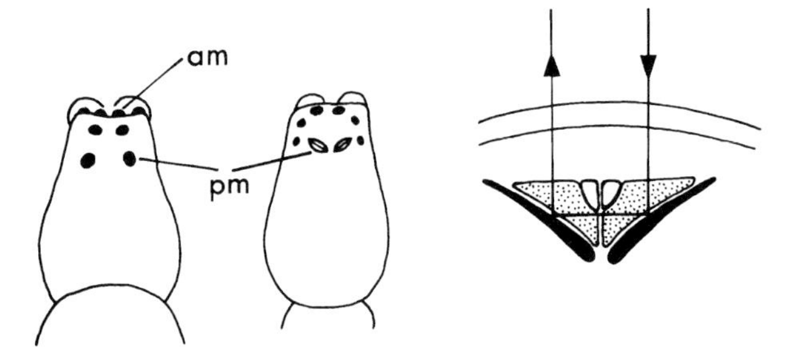

Figure 9.6. Left: the eyes of a wolf spider, *Arctosa* species, as seen from above; the forward-facing anterior median or 'principal' eyes (am) and upward facing posterior median (pm) eyes have been shown to be sensitive to the direction of polarisation. Centre: the 'squinty' boat-shaped polarisation sensitive posterior median (pm) eyes of the spider *Drassodes* as seen from above. Right: a section across a pm eye of *Drassodes*, with two reflecting layers (shown black) that lie obliquely below the masses of light-sensitive cells (dotted). The arrows show how light that is not absorbed on its first passage through one group of receptor cells is reflected back through them, across the eye and then through the other group of cells. Polarisation by reflection (chapter 7) acts together with the dichroicity of the receptor cells to enhance their polarisation sensitivity. There is no lens and no image is formed in these eyes but the two set at right angles are well adapted to detect the polarisation of sky light.

eyes have microvilli which, in parts of the retina, are arranged in orthogonal groupings. Electrical recordings have confirmed the presence of polarisation sensitivity in both wolf spiders and jumping spiders.

In 1999 a new kind of eye was discovered in another spider called *Drassodes*. Here again the two posterior median eyes are on top of the head, looking vertically upwards (figure 9.6). They are boat shaped and orientated at right angles to each other in a way that is aptly described as 'squinty'. In common with many other spider eyes, the retina is backed by a reflective layer which, as in the eyes of cats and many other nocturnal creatures, directs any unabsorbed light back through the retina. Such reflections from spiders' eyes can often be seen at night if one holds a torch close to one's face and looks around a garden lawn for instance. In *Drassodes* the reflection is bright blue and is polarised along the axis of the eye so that a rotating polariser extinguishes the reflection from each eye in turn. As shown in figure 9.6, the reflective layer is folded into two flat plates forming a V-shaped trench along the sides of the 'boat'. Light therefore reaches the retinal cells either directly or after reflection by one or both reflectors. Even an isolated piece of a reflector is found to polarise light and this clearly enhances the inherent dichroicity of the 60 or so main retinal cells whose microvilli run along the longer axis of the eye. The electrical responses of the cells peak in the ultraviolet and show an unusually high dependence on the direction of polarisation. These eyes have no lenses so they can form no image and they receive light from a wide angle, about 125°. The spiders are active around dawn and dusk when the sky overhead is polarised in a rather simple north–south band (figure 6.3 and chapter 6). Behavioural tests support the suggestion that the two eyes act together as a polarising sky compass to enable the spiders to return to their lairs after hunting forays. The structure of the polarising reflector layers is not yet known but it probably involves multiple layers and a Brewster-type reflection (chapter 7).

In the retinas of vertebrate eyes, the receptor cells, rods and cones, are quite different in structure from any of the foregoing. Instead of having microvilli, they have a stack of transverse disclike membranes that lie across the light path and contain their receptor molecules in random orientation (figure 9.7) in their walls. Indeed within the rather fluid membranes the individual molecules are free to rotate around the optical axis so that they all point randomly in different directions and fluctuate continuously. This arrangement would not be expected to be dichroic and optical measurements have confirmed this. In fact, when light is shone experimentally *across* the cell, there is marked dichroicity since all the receptor molecules are at least parallel with the planes of their holding membranes and strongly absorb when the direction of polarisation is in this plane. But in life and a whole eye, light never shines across the cells, only down their length.

The freedom of receptor molecules to rotate within the membranes of vertebrate retinal cells has been demonstrated by some ingenious observations. First, the visual cells of a frog were treated with a chemical (glutaraldehyde) that killed the cells and 'set' the fluid membranes, so preventing further movement of their constituents. A bright flash of polarised light shining in the normal direction along the length of the cells then left them strongly dichroic for further beams of light along this axis. Any receptor molecules that had absorbed light with the direction of polarisation of the first flash were inactivated (when they are 'bleached' and no longer absorb light), so further absorption in a

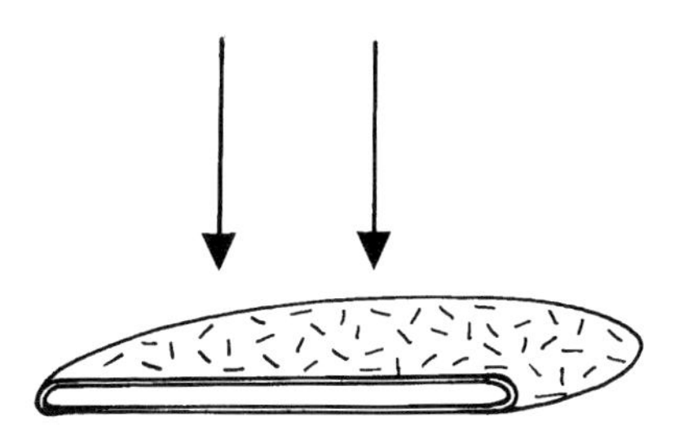

Figure 9.7. The light-sensitive cells, both rods and cones, of vertebrate eyes contain disclike double membranes that lie transversely to the light path (arrows). The light-sensitive molecules (dashes) are contained within these membranes but they point in all directions and are therefore not sensitive to the polarisation of light coming from the lens. This arrangement contrasts with the regularly orientated molecules in the eyes of bees and other invertebrates (figure 9.4).

second flash was favoured for light polarised at right angles to the first, in the molecules that were still receptive. But in living cells, where these molecules are free to rotate rapidly and randomly, such dichroicity persists for only a very brief time, around 20 μs (millionths of a second). Restoration of sensitivity by the regeneration of bleached molecules is very slow by comparison, taking a matter of minutes, and so cannot explain the rapid loss of dichroicity in normal eyes after the first flash. In this way paired flashes of polarised light can be used to measure the rate of molecular rotation within the membranes of living visual cells.

From all this one would not expect vertebrates to be sensitive to the direction of polarisation of light, and until a few years ago this was believed to be so. One exception may be found in the anchovy fish, *Anchoa* species, where the cone cells of the retina are contorted so that the membranous discs run almost lengthways and so lie edge on to the light path (figure 9.8). The cones are arranged in the retina in vertical rows (with rods in between) and there are two kinds of cone that alternate in each row, with the two types having their discs arranged orthogonally, either vertically or horizontally. This arrangement is expected to make each cone dichroic and the two orthogonal directions could give the anchovy the ability to detect the direction of polarisation. The way in which one kind of cone fits neatly under the other may increase overall sensitivity since the horizontally polarised light that passes through the short cones may then be absorbed by the horizontally sensitive long cones and not be wasted. In this respect there is a functional resemblence

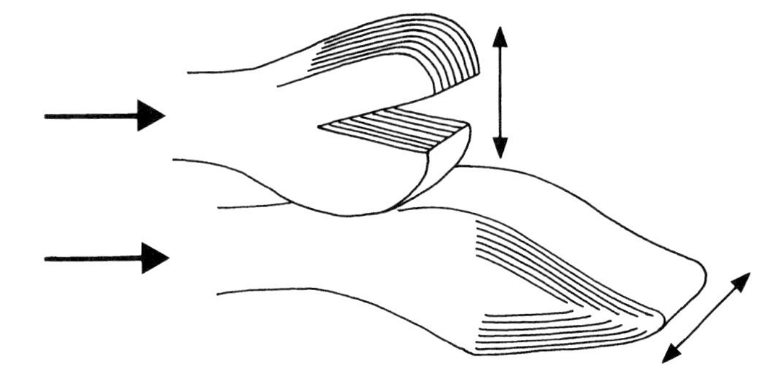

Figure 9.8. The outer, light-sensitive regions of retinal cone cells in the anchovy. The edges of the sensitive membranes are indicated by the close parallel lines (though there are very many more than shown here). The double-headed arrows show the planes of the membranes, which are tilted to lie along the normal light path and are therefore likely to be dichroic. There are two kinds of cell that alternate along the vertical rows of cones within the retina. The short forked cones have vertical membranes and should respond to vertically polarised light, while the long pointed cones have horizontal membranes and should respond to horizontally polarised light. This orthogonal arrangement is unknown among other vertebrates and probably forms a polarisation analyser. Light that has entered the eye through the lens comes from the left as shown by the larger arrows.

to the twisted arrangement of the cell bundles in insect eyes. So far this arrangement appears to be unique among vertebrates and the reason for it is unknown.

Another possibility involves the eyes of some fish such as trout, in which the retina retains a fold called the embryonic fissure into later life. In some cases this fold holds small numbers of cone cells that are therefore sideways on to the light path and would be expected to be dichroic. But no correlation has been attempted between this arrangement and the ability to detect polarisation, nor would it enable the fish to analyse more than a tiny part of the visual field.

Nevertheless evidence has steadily accumulated to suggest that many other non-mammalian vertebrates may be sensitive to polarisation and even observe its direction. For instance a variety of fish including trout and goldfish show spontaneous responses such as orientation to the direction of polarisation. Salamanders and pigeons can be trained to respond to the direction of polarisation; hatchling turtles seem to orientate towards the sea by observing polarisation; some small migratory birds appear to use the sky compass (chapter 6) after dusk, and during development may 'set' an internal sense of the earth's magnetic field by comparison with the polarised sky compass.

The mechanism is still not clear but it seems that polarisation dependence in these vertebrates may be associated with 'double cones' in the retina. These comprise two similar cone cells in an intimate association, the combined pair having an elliptical cross section. It has been suggested that these pairs might somehow conduct light in a birefringent way and therefore may be polarisation dependent. This suggestion is supported by the layout of the double cones in the retina, where they are set in a square mosaic of tetrads with the long axes of the ellipses set in two orthogonal directions. Furthermore the polarisation sensitivity of the Green Sunfish, *Lepomis cyanellus*, has been found to be best at the same red wavelength as light absorption by the double cones, whereas in this fish no other cones have the same pigment. It also seems to be significant that double cones are found in fishes, amphibia, reptiles and birds but not in mammals, where no case of polarisation sensitivity has (yet) been discovered.

The story is far from clear, however. Electrical recordings from single cells and nerve fibres in trout and goldfish have shown that green and red double cones respond to the same direction of polarisation although they are orientated at right angles to each other. A common basis of birefringence due to asymmetry of cross section seems unlikely therefore. Also the ultraviolet cones (but not the blue ones) respond strongly to polarisation direction although they are not formed into double cone pairs.

Finally, there is one way in which humans are sometimes said to be able to detect strong polarisation directly, without reference to reflections or scattering. The effect is not strong and I myself, in common with others, have great difficulty seeing it. By gazing at an even field of strongly polarised light, especially if it is blue in colour (such as a clear sky at right angles to the sun) one is supposed to see a small faint figure, called Haidinger's brush after its discoverer who described it in 1844. It consists (figure 9.9) of two yellow, brushlike patterns back to back and with blue areas between. The whole figure is about 3° wide (the full moon is about half a degree wide) and the yellow–brown wings are aligned at right angles to the direction of polarisation. I have only seen it very faintly after staring for some time with one eye through a

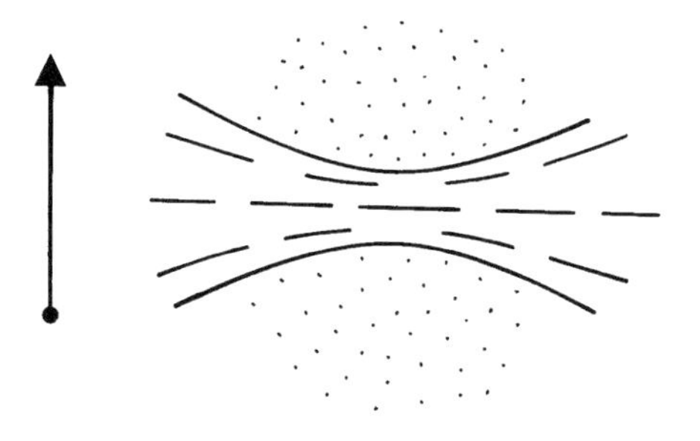

Figure 9.9. An impression of the pattern known as Haidinger's brushes that is sometimes seen when people look at a patch of polarised light (from various sources). The barlike figure (dashes) with open fuzzy ends, the 'brushes', is faintly yellow–brown while the (dotted) areas either side of its waist are bluish. The axis of the figure is at right angles to the direction of polarisation, shown by the arrow, and extends for about 3°.

polaroid film at a brightly lit sheet of white paper. Even then it only works for me when the polaroid is suddenly rotated by 90° or so and the figure appears as it rotates too. It is also said that circularly polarised light can produce brushes, aligned lower left to upper right for right-hand circular polarisation and upper left to lower right for left-hand circular polarisation. I cannot vouch for this.

There is no consensus on the explanation for these figures although they are commonly attributed rather vaguely to 'dichroism in the fovea', or that part of the retina responsible for the most detailed central vision and sometimes called the yellow spot. The usefulness of the figures is probably negligible and they are mentioned here simply because some people say that humans *can* see the effects of polarisation after all. To see a small figure, however, is not at all the same as being able to see the property of polarisation in an image. What the world might actually look like to animals which have that ability is, after all, something we can only imagine. Our knowledge of the various phenomena described in this book is only the beginning. We can only dream about its part in any whole sensory experience.

Some recommendations for further reading

Bunn C 1964 *Crystals: Their Role in Nature and in Science* (New York: Academic)

Greenler R 1980 *Rainbows, Halos and Glories* (Cambridge: Cambridge University Press) (paperback edn 1989).

Gribble C D and Hall A J 1985 *A Practical Introduction to Optical Mineralogy* (London: Allen and Unwin) (paperback)

Hartshorne N H and Stuart A 1960 *Crystals and the Polarising Microscope* 3rd edn (London: Arnold)

Konnen G P 1985 *Polarised Light in Nature* (Cambridge: Cambridge University Press) (first published 1980 in Dutch by Thieme & Cie-Zutphen, The Netherlands)

Land E H 1951 Some aspects of the development of sheet polarizers *J. Opt. Soc. Am.* **41** 957–63

Lowry T M 1964 *Optical Rotatory Power* (New York: Dover) (1st edn 1935 (London: Longmans Green))

Lynch D K and Livingston W 1995 *Color and Light in Nature* (Cambridge: Cambridge University Press)

Lythgoe J N 1979 *The Ecology of Vision* (Oxford: Clarendon)

Minnaert M 1940 *The Nature of Light and Colour in the Open Air* (London: Bell) (paperback edn 1954 (New York: Dover))

Robinson P C and Bradbury S 1992 *Qualitative Polarized-Light Microscopy (Royal Microscopical Society Microscopy Handbook 9)* (Oxford: Oxford University Press)

Shurcliff W A and Ballard S S 1964 *Polarized Light* (Princeton, NJ: Van Nostrand) (paperback)

Smith H G 1956 *Minerals and the Microscope* 4th edn, revised by M K Wells (London: Murby) (paperback)

Spottiswoode W 1874 *Polarisation of Light* (London: Macmillan) van de Hulst H C 1957 *Light Scattering by Small Particles* (New York: Wiley) (paperback edition 1981 (New York: Dover))

von Frisch K 1950 *Bees: Their Vision, Chemical Senses and Language* (Ithaca, NY: Cornell University Press) (paperback and later editions in UK)

——1954 *The Dancing Bees* (London: Methuen) (University Paperback edition 1970)

Waterman T H 1981 Polarization sensitivity in *Handbook of Sensory Physiology vol VII/6B, Comparative Physiology and Evolution of Vision in Invertebrates, B: Invertebrate Visual Centers and Behavior I* ed H Autrum (Berlin: Springer) pp 281–469

Wehner R 1987 'Matched filters'—neural models of the external world *J. Comparative Physiol.* A **161** 511–31

Wood E A 1964 *Crystals and Light: An Introduction to Optical Crystallography* (Princeton, NJ: Van Nostrand) (revised paperback edn 1977 (New York: Dover))

Polaroid materials of various kinds are available from many optical and general scientific suppliers. I am especially grateful for helpful advice, information and small quantities of unusual products that have been supplied by Optical Filters Ltd, The Business Centre, 14 Bertie Road, Thame, Oxon OX9 3FR, UK (01844-260377).

Index

Index of names

Subject index